HYPERSPHERICAL HARMONICS AND GENERALIZED STURMIANS

Progress in Theoretical Chemistry and Physics

VOLUME 4

Editor-in-Chief:

J. Maruani *(Laboratoire de Chimie Physique, Paris, France)*
S. Wilson *(Rutherford Appleton Laboratory, Oxfordshire, United Kingdom)*

Editorial Board:

Hyperspherical Harmonics and Generalized Sturmians

by

John Avery

H.C. Ørsted Institute,
University of Copenhagen,
Copenhagen, Denmark

KLUWER ACADEMIC PUBLISHERS

DORDRECHT / BOSTON / LONDON

A C.I.P. Catalogue record for this book is available from the Library of Congress.

ISBN 1-4020-0409-5
Transferred to Digital Print 2001

Published by Kluwer Academic Publishers,
P.O. Box 17, 3300 AA Dordrecht, The Netherlands.

Sold and distributed in North, Central and South America
by Kluwer Academic Publishers,
101 Philip Drive, Norwell, MA 02061, U.S.A.

In all other countries, sold and distributed
by Kluwer Academic Publishers,
P.O. Box 322, 3300 AH Dordrecht, The Netherlands.

Printed on acid-free paper

Dedicated to Professors
Dudley R. Herschbach
Carl E. Wulfman and
Vincenzo Aquilanti
with admiration for their
pioneering research.

Contents

Introduction

This book explores the connections between the theory of hyperspherical harmonics, momentum-space quantum theory, and generalized Sturmian basis functions; and it introduces methods which may be used to solve many-electron problems directly, without the use of the self-consistent-field approximation.

Among the pioneers of momentum-space quantum theory, the names of Podolski, Pauling, Fock, Coulson, McWeeny, Duncanson, Shibuya, Wulfman, Judd, Koga and Aquilanti are outstanding. Fourier transformed hydrogenlike wave functions were obtained in 1928 by Podolski and Pauling [241, 242]. Later, Fock [124, 125, 47] was able to solve the Schrödinger equation for hydrogenlike atoms directly in momentum space. In a remarkably brilliant paper, Fock introduced a transformation which maps 3-dimensional momentum space onto the surface of a 4-dimensional unit hypersphere; and he was able to show that when this transformation is made, the momentum-space hydrogenlike orbitals can be very simply expressed in terms of 4-dimensional hyperspherical harmonics. Fock also demonstrated that the unexpected n^2-fold degeneracy of hydrogenlike wave functions corresponds to the number of linearly independent hyperspherical harmonics belonging to a particular value of the grand angular momentum quantum number, $\lambda = n - 1$.

In the early 1940's, Coulson, McWeeny and Duncanson [72, 73, 108, 217, 218] explored the possibilities for solving the momentum space Schrödinger equation for more complicated atoms and molecules. They found that although interesting insights could be gained through the momentum-space approach, some of the necessary integrals were very difficult; and at the time, these difficult integrals presented a barrier to further progress.

In 1965, Shibuya and Wulfman [272] published a pioneering paper in which they extended Fock's momentum-space method to the problem of a single particle moving in a many-center Coulomb potential. Shibuya and Wulfman evaluated the integrals needed for the calculation by using the Wigner coefficients of 4-dimensional hyperspherical harmonics. Wulfman [316-318] was able to show how this new method was connected with dynamical symmetry and with Lie groups.

In 1975, the first book on the momentum-space methods of Fock, Shibuya and Wulfman was published by Judd [170]. This book, entitled *Angular Momentum Theory for Diatomic Molecules*, develops in a systematic way the theory of 4-dimensional hyperspherical harmonics and their application to the problem of an electron moving in a many-center Coulomb potential. The application of the method to diatomic molecules is discussed in detail in Judd's excellent book.

Further progress in the development of the Shibuya-Wulfman method was made by Monkhost and Jeziorski [227], by Dachon, Dumont-Lepage and Gazeau [105, 106], by Novosadov [232, 233], and especially by Koga and his co-workers in Japan [186-195]. By using a secular equation based on the second-iterated momentum-space Schrödinger equation, Koga and Matsuhashi [190-192] were able to obtain the electronic energy of H_2^+ with 10-figure accuracy as a function of the internuclear separation R. Their momentum-space results agree with the best available direct-space calculations, but are of higher precision for low values of R.

Other important contributions to the development of the Shibuya-Wulfman method were made by Aquilanti and his coworkers at the University of Perugia in Italy [9-28, 70]. These authors demonstrated that alternative sets of hyperspherical harmonics may be better adapted to particular physical problems, such as calculations of the Stark effect or the quadratic Zeeman effect, or calculations on diatomic molecules. The *method of trees*, which had been introduced into nuclear physics by a number of Russian authors [176, 177, 184, 202, 230, 277, 282, 296, 297] was further developed by Aquilanti and his group, and was applied by them to the generation of alternative orthonormal sets of hyperspherical harmonics. These authors discussed in detail an alternative set of hyperspherical harmonics which, under Fock's mapping, represent the momentum-space counterpart of direct-space solutions in parabolic coordinates; and they were able to find explicit expressions for the generalized Clebsch-Gordan coefficients relating this alternative set to standard hyperspherical harmonics.

Sturmian basis functions have an independent history, although they are closely related to the topics which we have just been discussing. Very early in the history of quantum theory, it was thought that hydrogenlike wave functions could be used as building blocks to construct the wave functions of more complicated systems - for example many-electron atoms or molecules. However, it was soon realized that unless the continuum is included, a set of hydrogenlike orbitals is not complete. To remedy this defect, Shull and Löwdin [273] introduced sets of radial functions which could be expressed in terms of Laguerre polynomials multiplied by exponential factors. The sets were constructed in such a way as to be complete, i.e. any radial function obeying the appropriate boundary conditions could be expanded in terms of the Shull-Löwdin basis sets. Later Rotenberg [256, 257] gave the name "Sturmian" to basis sets of this type in order to emphasize their connection with Sturm-Liouville theory. There is a large and rapidly-growing literature on Sturmian basis functions; and selections from this literature are cited in the bibliography.

In 1968, Goscinski [138] completed a study of the properties of Sturmian basis sets, formulating the problem in such a way as to make generalization of the concept very easy. In the present text, we shall follow Goscinski's easily generalizable definition of Sturmians.

Generalized Sturmian basis sets can be understood from the following simple argument: Suppose that we are able to find a set of solutions to the wave equation

$$\left[-\frac{1}{2}\Delta + \beta_\nu V_0(\mathbf{x}) - E \right] \phi_\nu(\mathbf{x}) = 0$$

Here

$$\Delta \equiv \sum_{j=1}^{d} \frac{\partial^2}{\partial x_j^2}$$

is the generalized Laplacian operator and

$$\mathbf{x} = \{x_1, x_2, ..., x_d\}$$

are the mass-weighted Cartesian coordinates of the particles in a system. If we are discussing a single particle, then $d = 3$, while if we are discussing an N-particle system, $d = 3N$. We let β_ν be a set of constants chosen in such a way that the functions $\phi_\nu(\mathbf{x})$ all correspond to the same value of energy, E, regardless of their quantum numbers. Then it is easy to show, using the Hermiticity of the operator $\frac{1}{2}\Delta + E$, that the functions in the set obey a potential-weighted orthonormality relation:

$$\int dx \phi_{\nu'}^\dagger(\mathbf{x}) V_0(\mathbf{x}) \phi_\nu(\mathbf{x}) = \delta_{\nu',\nu} \frac{2E}{\beta_\nu}$$

The normalization of the functions in the generalized Sturmian basis set can, of course, be chosen according to convenience. We shall see that the choice of normalization shown above is a natural and convenient one, and that it is closely related to the orthonormality of the hyperspherical harmonics.

Now suppose that we wish to solve a Schrödinger equation of the form

$$\left[-\frac{1}{2}\Delta + V(\mathbf{x}) - E \right] \psi(\mathbf{x}) = 0$$

If we expand the wave function $\psi(\mathbf{x})$ in terms of our generalized Sturmian basis set, we have

$$\sum_\nu \left[-\frac{1}{2}\Delta + V(\mathbf{x}) - E \right] \phi_\nu(\mathbf{x}) B_\nu = \sum_\nu \left[-\beta_\nu V_0(\mathbf{x}) + V(\mathbf{x}) \right] \phi_\nu(\mathbf{x}) B_\nu = 0$$

We next multiply from the left by an adjoint function in the set and integrate over coordinates, making use of the potential-weighted orthonormality relation. This leads to the secular equation:

$$\sum_\nu \left[\int dx \phi_{\nu'}^\dagger(\mathbf{x}) V(\mathbf{x}) \phi_\nu(\mathbf{x}) - 2E\delta_{\nu',\nu} \right] B_\nu = 0$$

Notice that the kinetic energy term has disappeared from the secular equation! This remarkable result always occurs when Sturmian basis functions are used.

The method just described is a very general one, since $V_0(\mathbf{x})$ may be chosen as a matter of convenience. Gazeau and Maquet [132, 133] introduced 2-electron Sturmians to study bound-state problems, while Bang, Vaagen and their coworkers [48, 49] pioneered the application of generalized Sturmians to problems in nuclear physics. Herschbach, Avery and Antonsen [30, 33, 35, 155] introduced and studied many-particle Sturmian basis sets where the "basis potential" $V_0(\mathbf{x})$ was taken to be the potential of a d-dimensional hydrogenlike atom; but in most applications, convergence with this basis proved to be slow. More recently, Aquilanti and Avery [28, 41, 44, 45] introduced many-electron

Sturmian basis sets where $V_0(\mathbf{x})$ was taken to be the actual external potential experienced by the N electrons in an atom, i.e., the attractive Coulomb potential of the nucleus; and convergence proved to be very rapid. The method was extended to molecules by Avery and Sauer [41, 45], with the help of Fernández Rico and López [121, 122, 212], using the Shibuya-Wulfman technique for constructing many-electron molecular Sturmians.

I hope that you will enjoy reading this book and that you will find that the method of many-electron Sturmians offers an interesting and fresh alternative to the usual SCF-CI methods for calculating atomic and molecular structure. When many-electron Sturmians are used, and when the basis potential $V_0(\mathbf{x})$ is chosen to be the attractive potential of the nuclei in the system, the following advantages are offered:

1. The matrix representation of the nuclear attraction potential is diagonal.

2. The kinetic energy term vanishes from the secular equation.

3. The Slater exponents of the atomic orbitals are automatically optimized.

4. Convergence is rapid.

5. A correlated solution to the many-electron problem can be obtained directly, without the use of the SCF approximation.

6. Excited states can be obtained with good accuracy.

Acknowledgements

I am extremely grateful to the Carlsberg Foundation for travel grants which allowed me to make several visits to the laboratory of Prof. Dudley Herschbach at Harvard University. I would also like to thank the Danish Science Research Council which made possible two periods of research at the laboratory of Prof. Vincenzo Aquilanti at the University of Perugia, Italy, and also a short visit to the laboratory of Prof. Jamie Fernández Rico in Madrid. Prof. Fernández Rico and Dr. Rafael López have generously allowed me to use their computer program for evaluating many-center two-electron integrals over Slater-type orbitals. It has been an extremely great pleasure for me to receive encouragement from two of the most important pioneers of momentum-space quantum theory - Prof. Roy McWeeny and Prof. Carl Wulfman. I am also grateful to Professors Mario Raimondi, Robert Nesbet, Luis Tel, Jens Bang and Jan Vaagen for encouragement and for stimulating discussions. Finally, I would like to thank my co-workers, Dr. Frank Antonsen, Dr. Wensheng Bian, Prof. John Loeser, Cand. Scient. Tom Børsen Hansen, Stud. Scient. Rune Shim, Dr. Cecilia Coletti, and Assist. Prof. Stephan Sauer for their important contributions.

Chapter 1

MANY-PARTICLE STURMIANS

Suppose that we are able to find a set of functions $\phi_\nu(\mathbf{x})$ which are solutions to the many-electron equation:

$$\left[-\frac{1}{2}\Delta + \beta_\nu V_0(\mathbf{x}) - E \right] \phi_\nu(\mathbf{x}) = 0 \qquad (1.1)$$

In equation (1.1), Δ is the generalized Laplacian operator

$$\Delta \equiv \sum_{j=1}^{N} \left(\frac{\partial^2}{\partial x_j^2} + \frac{\partial^2}{\partial y_j^2} + \frac{\partial^2}{\partial z_j^2} \right) \equiv \sum_{j=1}^{N} \Delta_j \qquad (1.2)$$

while $\mathbf{x}$ represents the set of coordinate vectors for all the electrons in the system:

$$\mathbf{x} = \{\mathbf{x}_1, \mathbf{x}_2, ..., \mathbf{x}_N\} \qquad (1.3)$$

and

$$\mathbf{x}_j \equiv \{x_j, y_j, z_j\} \qquad j = 1, 2, ..., N \qquad (1.4)$$

The potential, $V_0(\mathbf{x})$, is chosen to be the Coulomb attraction potential due to the nuclei in the system. The simplest case is that of an N-electron atom or ion, where

$$V_0(\mathbf{x}) = -\sum_{j=1}^{N} \frac{Z}{r_j} \qquad (1.5)$$

7

represents the attractive potential of a single nucleus. Let us begin with this simple case, and later consider molecules, where $V_0(\mathbf{x})$ represents the attractive potential of all the nuclei in the system. If the constants β_ν in equation (1.1) are especially chosen so that all the solutions to the zeroth-order many-electron wave equation (1.1) correspond to the same value of the energy E, regardless of the quantum numbers ν, then the set of functions $\phi_\nu(\mathbf{x})$ may be regarded as a set of generalized Sturmian basis functions.

Is it really possible to obtain a set of solutions to equation (1.1)? And is it possible to find a suitable set of weighting factors for the potential so that all these solutions correspond to the same energy? We shall now try to demonstrate that such a set of solutions can indeed be constructed. Let

$$\phi_\nu(\mathbf{x}) = \chi_\mu(\mathbf{x}_1)\chi_{\mu'}(\mathbf{x}_2)\chi_{\mu''}(\mathbf{x}_3)... \tag{1.6}$$

where

$$\begin{aligned}
\chi_{nlm,+1/2}(\mathbf{x}_j) &= R_{nl}(r_j)Y_{lm}(\theta_j,\phi_j)\alpha(j) \\
\chi_{nlm,-1/2}(\mathbf{x}_j) &= R_{nl}(r_j)Y_{lm}(\theta_j,\phi_j)\beta(j)
\end{aligned} \tag{1.7}$$

and

$$\begin{aligned}
R_{nl}(r_j) &= \mathcal{N}_{nl}(2k_\mu r_j)^l e^{-k_\mu r_j} F(l+1-n|2l+2|2k_\mu r_j) \\
\mathcal{N}_{nl} &= \frac{2k_\mu^{3/2}}{(2l+1)!}\sqrt{\frac{(l+n)!}{n(n-l-1)!}}
\end{aligned} \tag{1.8}$$

The functions $\chi_\mu(\mathbf{x}_j)$ are just the familiar hydrogenlike atomic spin-orbitals, except that the orbital exponents k_μ are left as free parameters which we shall determine later by means of subsidiary conditions. For example,

$$R_{1,0}(r) = 2k_\mu^{3/2}e^{-k_\mu r} \tag{1.9}$$

$$R_{2,1}(r) = 2k_\mu^{3/2}\frac{k_\mu r}{\sqrt{3}}e^{-k_\mu r} \tag{1.10}$$

and so on. In equation (1.6), μ represents the set of quantum numbers $\{n, l, m, s\}$. The functions $\chi_\mu(\mathbf{x}_j)$ satisfy the relationships:

$$\left[-\frac{1}{2}\Delta_j + \frac{1}{2}k_\mu^2 - \frac{nk_\mu}{r_j} \right] \chi_\mu(\mathbf{x}_j) = 0 \tag{1.11}$$

$$\int d\tau_j |\chi_\mu(\mathbf{x}_j)|^2 \frac{1}{r_j} = \frac{k_\mu}{n} \tag{1.12}$$

and

$$\int d\tau_j |\chi_\mu(\mathbf{x}_j)|^2 = 1 \tag{1.13}$$

We now introduce the subsidiary conditions

$$k_\mu^2 + k_{\mu'}^2 + k_{\mu''}^2 + \ldots = -2E \tag{1.14}$$

and

$$nk_\mu = n'k_{\mu'} = n''k_{\mu''} = \ldots = Z\beta_\nu \tag{1.15}$$

Provided that the subsidiary conditions are satisfied, the product shown in equation (1.6) will be a solution to (1.1), since

$$
\begin{aligned}
\frac{1}{2}\Delta\phi_\nu(\mathbf{x}) &= \left[\frac{1}{2}\Delta_1 + \frac{1}{2}\Delta_2 + \ldots \right] \chi_\mu(\mathbf{x}_1)\chi_\mu(\mathbf{x}_2)\ldots \\
&= \left[\frac{1}{2}k_\mu^2 - \frac{nk_\mu}{r_1} + \frac{1}{2}k_{\mu'}^2 - \frac{n'k_{\mu'}}{r_2} + \ldots \right] \chi_\mu(\mathbf{x}_1)\chi_{\mu'}(\mathbf{x}_2)\ldots \\
&= \left[-E - \frac{\beta_\nu Z}{r_1} - \frac{\beta_\nu Z}{r_2} - \ldots \right] \chi_\mu(\mathbf{x}_1)\chi_{\mu'}(\mathbf{x}_2)\ldots \\
&= \left[-E + \beta_\nu V_0(\mathbf{x}) \right] \phi_\nu(\mathbf{x}) \tag{1.16}
\end{aligned}
$$

It can easily be seen that an antisymmetrized function of the form

$$
\begin{aligned}
\phi_\nu(\mathbf{x}) &\equiv |\chi_\mu \chi_{\mu'} \chi_{\mu''} \ldots| \\
&\equiv \frac{1}{\sqrt{N!}} \begin{vmatrix} \chi_\mu(\mathbf{x}_1) & \chi_{\mu'}(\mathbf{x}_1) & \chi_{\mu''}(\mathbf{x}_1) & \cdots \\ \chi_\mu(\mathbf{x}_2) & \chi_{\mu'}(\mathbf{x}_2) & \chi_{\mu''}(\mathbf{x}_2) & \cdots \\ \vdots & \vdots & \vdots & \vdots \\ \chi_\mu(\mathbf{x}_N) & \chi_{\mu'}(\mathbf{x}_N) & \chi_{\mu''}(\mathbf{x}_N) & \cdots \end{vmatrix} \tag{1.17}
\end{aligned}
$$

will also satisfy (1.1), since the antisymmetrized function can be expressed as a sum of terms, each of which has the form shown in equation

(1.6). When we use the set of functions $\phi_\nu(\mathbf{x})$ as a basis set to build up the wave function of an N-electron atom or ion, we shall of course use the antisymmetrized functions shown in equation (1.17), since we wish the wave function to satisfy the Pauli principle.

Sturmian basis sets satisfy potential-weighted orthonormality relations, and similarly, our generalized many-electron Sturmian basis functions satisfy an orthonormality relation where the weighting factor is the potential $V_0(\mathbf{x})$. To see this, we consider two different solutions to equation (1.1). From (1.1), it follows that they satisfy

$$\int dx \phi_{\nu'}^*(\mathbf{x}) \left[\frac{1}{2}\Delta + E\right] \phi_\nu(\mathbf{x}) = \beta_\nu \int dx \phi_{\nu'}^*(\mathbf{x}) V_0(\mathbf{x}) \phi_\nu(\mathbf{x}) \qquad (1.18)$$

and

$$\int dx \phi_\nu^*(\mathbf{x}) \left[\frac{1}{2}\Delta + E\right] \phi_{\nu'}(\mathbf{x}) = \beta_{\nu'} \int dx \phi_\nu^*(\mathbf{x}) V_0(\mathbf{x}) \phi_{\nu'}(\mathbf{x}) \qquad (1.19)$$

If we take the complex conjugate of (1.19), subtract it from (1.18), and make use of the Hermiticity of the operator $\frac{1}{2}\Delta + E$, we obtain:

$$(\beta_\nu - \beta_{\nu'}) \int dx \phi_{\nu'}^*(\mathbf{x}) V_0(\mathbf{x}) \phi_\nu(\mathbf{x}) = 0 \qquad (1.20)$$

where the constants β_ν are assumed to be real. From equation (1.20) it follows that if $\beta_\nu - \beta_{\nu'} \neq 0$, then

$$\int dx \phi_{\nu'}^*(\mathbf{x}) V_0(\mathbf{x}) \phi_\nu(\mathbf{x}) = 0 \qquad (1.21)$$

When $\nu = \nu'$, we have (from (1.12), (1.14) and (1.15))

$$\begin{aligned}
\int dx V_0(\mathbf{x}) |\phi_\nu(\mathbf{x})|^2 &= -\sum_{\mu \subset \nu} Z \int d\tau_j |\chi_\mu(\mathbf{x}_j)|^2 \frac{1}{r_j} \\
&= -\sum_{\mu \subset \nu} Z \frac{k_\mu}{n} \\
&= -\frac{1}{\beta_\nu} \sum_{\mu \subset \nu} k_\mu^2 = \frac{2E}{\beta_\nu} \qquad (1.22)
\end{aligned}$$

Combining (1.21) and (1.22), and making use of the orthonormality of the spin functions and the spherical harmonics, we obtain the potential-weighted orthonormality relations:

$$\int dx \; \phi_{\nu'}^*(\mathbf{x}) V_0(\mathbf{x}) \phi_\nu(\mathbf{x}) = \delta_{\nu',\nu} \frac{2E}{\beta_\nu} \tag{1.23}$$

where ν stands for the set of quantum numbers $\{\mu, \mu', \mu'', ...\}$.

The secular equation

Having constructed a set of many-electron Sturmian basis functions by the method just described, we would like to use this basis set to solve the Schrödinger equation for an N-electron atom. Using atomic units, we can write the Schrödinger equation in the form:

$$\left[-\frac{1}{2}\Delta + V(\mathbf{x}) - E \right] \psi(\mathbf{x}) = 0 \tag{1.24}$$

where

$$V(\mathbf{x}) = V_0(\mathbf{x}) + V'(\mathbf{x}) \tag{1.25}$$

Here $V_0(\mathbf{x})$ is the nuclear attraction potential shown in equation (1.5), while

$$V'(\mathbf{x}) = \sum_{i>j}^{N} \sum_{j=1}^{N} \frac{1}{r_{ij}} \tag{1.26}$$

is the interelectron repulsion potential. Expanding the wave function as series of generalized Sturmian basis functions, and making use of equation (1.1), we obtain:

$$\sum_\nu \left[-\frac{1}{2}\Delta + V(\mathbf{x}) - E \right] \phi_\nu(\mathbf{x}) B_\nu = \sum_\nu \left[-\beta_\nu V_0(\mathbf{x}) + V(\mathbf{x}) \right] \phi_\nu(\mathbf{x}) B_\nu = 0 \tag{1.27}$$

Multiplying (1.27) from the left by a conjugate basis function, integrating over the coordinates, and making use of the potential-weighted orthonormality relation, (1.23), we obtain:

$$\sum_\nu \left[\int dx \phi_{\nu'}^*(\mathbf{x}) V(\mathbf{x}) \phi_\nu(\mathbf{x}) - 2E \delta_{\nu',\nu} \right] B_\nu = 0 \tag{1.28}$$

We now introduce a parameter, p_0, which is related to the electronic energy E of the system by

$$p_0^2 \equiv -2E \qquad (1.29)$$

and we also introduce the definition:

$$T_{\nu',\nu} \equiv -\frac{1}{p_0} \int dx \phi_{\nu'}^*(\mathbf{x}) V(\mathbf{x}) \phi_\nu(\mathbf{x}) \qquad (1.30)$$

We shall see below that if $V(\mathbf{x})$ is a potential produced by Coulomb interactions, then the matrix $T_{\nu',\nu}$, defined in this way, is independent of p_0. In terms of these new parameters, the secular equation, (1.28), can be written in the form:

$$\sum_\nu \left[T_{\nu',\nu} - p_0 \delta_{\nu',\nu}\right] B_\nu = 0 \qquad (1.31)$$

The total potential, V, consists of a nuclear attraction part, V_0, and an interelectron repulsion part, V'. The matrix $T_{\nu',\nu}$ can also be divided into two parts

$$T_{\nu',\nu} = T_{\nu',\nu}^0 + T_{\nu',\nu}' \qquad (1.32)$$

corresponding respectively to nuclear attraction and interelectron repulsion. From the potential-weighted orthonormality relation (1.23) and the subsidiary conditions, (1.14) and (1.15), it follows that:

$$
\begin{aligned}
T_{\nu',\nu}^0 &= -\frac{1}{p_0} \int dx \phi_{\nu'}^*(\mathbf{x}) V_0(\mathbf{x}) \phi_\nu(\mathbf{x}) \\
&= -\frac{2E}{p_0 \beta_\nu} \delta_{\nu',\nu} = \frac{p_0}{\beta_\nu} \delta_{\nu',\nu} = \frac{1}{\beta_\nu} \left(\sum_{\mu \subset \nu} k_\mu^2\right)^{1/2} \delta_{\nu',\nu} \\
&= Z \left(\sum_{\mu \subset \nu} \frac{1}{n^2}\right)^{1/2} \delta_{\nu',\nu} \qquad (1.33)
\end{aligned}
$$

Thus $T_{\nu',\nu}^0$ is independent of p_0, as we mentioned above. If interelectron repulsion is neglected, the matrix $T_{\nu',\nu} \approx T_{\nu',\nu}^0$ is diagonal, and the Sturmian secular equation, (1.31), simply requires that

$$p_0 \approx Z \left(\sum_{\mu \subset \nu} \frac{1}{n^2}\right)^{1/2} \qquad (1.34)$$

The zeroth-order energy of the system is then given by

$$E = -\frac{p_0^2}{2} \approx -\frac{Z^2}{2} \sum_{\mu \subset \nu} \frac{1}{n^2} \tag{1.35}$$

which is the correct energy of a system of N noninteracting electrons in the attractive field of the atom's nucleus. (For simplicity we use the approximation where the motion of the nucleus is neglected.)

The interelectron repulsion matrix

Let us now turn to the evaluation of the interelectron repulsion matrix,

$$T'_{\nu',\nu} \equiv -\frac{1}{p_0} \int dx \phi^*_{\nu'}(\mathbf{x}) \sum_{i>j}^{N} \sum_{j=1}^{N} \frac{1}{r_{ij}} \phi_\nu(\mathbf{x}) \tag{1.36}$$

We shall see that this matrix is also independent of p_0. In order to evaluate $T'_{\nu',\nu}$, we need to calculate 2-electron integrals of the form:

$$\begin{aligned} J &= \int_0^\infty dr_1 r_1^{2+j_1} e^{-\zeta_1 r_1} \int_0^\infty dr_2 r_2^{2+j_2} e^{-\zeta_2 r_2} \\ &\times \int d\Omega_1 W_1(\hat{\mathbf{x}}_1) \int d\Omega_2 W_2(\hat{\mathbf{x}}_2) \frac{1}{r_{12}} \end{aligned} \tag{1.37}$$

where $W_1(\hat{\mathbf{x}}_1)$ and $W_2(\hat{\mathbf{x}}_2)$ are products of spherical harmonics. We can expand $1/r_{12}$ in a series of Legendre polynomials:

$$\frac{1}{r_{12}} = \sum_{l=0}^{\infty} \frac{r_<^l}{r_>^{l+1}} P_l(\hat{\mathbf{x}}_1 \cdot \hat{\mathbf{x}}_2) \tag{1.38}$$

Then

$$J = \sum_{l=0}^{\infty} a_l I_l \tag{1.39}$$

where

$$a_l \equiv \int d\Omega_1 W_1(\hat{\mathbf{x}}_1) \int d\Omega_2 W_2(\hat{\mathbf{x}}_2) P_l(\hat{\mathbf{x}}_1 \cdot \hat{\mathbf{x}}_2) \tag{1.40}$$

and

$$I_l \equiv \int_0^\infty dr_1 r_1^{j_1+2} e^{-\zeta_1 r_1} \int_0^\infty dr_2 r_2^{j_2+2} e^{-\zeta_2 r_2} \frac{r_<^l}{r_>^{l+1}} \tag{1.41}$$

Usually very few of the angular coefficients a_l are non-zero, so the sum in equation (1.39) involves only a few terms. The radial integrals I_l can be evaluated by means of the relationship (Appendix B):

$$\int_0^\infty dr_1 \, r_1^{j_1+2} e^{-\zeta_1 r_1} \int_0^\infty dr_2 \, r_2^{j_2+2} e^{-\zeta_2 r_2} \frac{r_<^l}{r_>^{l+1}}$$

$$= \frac{\Gamma(j_1 + j_2 + 5)}{(\zeta_1 + \zeta_2)^{j_1+j_2+4}} \left[\frac{{}_2F_1(1, l - j_1 - 1; j_2 + l + 4; -\zeta_2/\zeta_1)}{(j_2 + l + 3)\zeta_1} \right.$$

$$\left. + \frac{{}_2F_1(1, l - j_2 - 1; j_1 + l + 4; -\zeta_1/\zeta_2)}{(j_1 + l + 3)\zeta_2} \right] \tag{1.42}$$

where

$${}_2F_1(a, b; c; x) \equiv 1 + \frac{ab}{c}x + \frac{a(a+1)b(b+1)}{c(c+1)} \frac{x^2}{2!} + \ldots \tag{1.43}$$

is a hypergeometric function. For example, suppose that we are considering a 2-electron atom or ion and that

$$\phi_\nu(\mathbf{x}) = |\chi_{1s}\chi_{\bar{1}s}| \equiv \frac{1}{\sqrt{2}}[\chi_{1s}(1)\chi_{\bar{1}s}(2) - \chi_{1s}(2)\chi_{\bar{1}s}(1)] \tag{1.44}$$

where

$$\chi_{1s}(1) = \left(\frac{k_\mu^3}{\pi}\right)^{1/2} e^{-k_\mu r_1} \alpha(1)$$

$$\chi_{\bar{1}s}(1) = \left(\frac{k_\mu^3}{\pi}\right)^{1/2} e^{-k_\mu r_1} \beta(1) \tag{1.45}$$

so that

$$\phi_\nu(\mathbf{x}) = \frac{k_\mu^3}{\pi} e^{-k_\mu(r_1+r_2)} \frac{1}{\sqrt{2}} [\alpha(1)\beta(2) - \beta(1)\alpha(2)] \tag{1.46}$$

Then the diagonal element of the interelectron repulsion matrix involving the configuration $\nu = \{1, 0, 0, \frac{1}{2}; 1, 0, 0, -\frac{1}{2}\}$ is given by

$$T'_{\nu,\nu} = -\frac{1}{p_0} \int d\mathbf{x} \phi_\nu^*(\mathbf{x}) \frac{1}{r_{12}} \phi_\nu(\mathbf{x})$$

$$= -\frac{1}{p_0}\left(\frac{k_\mu^3}{\pi}\right)^2 \int d^3x_1 \int d^3x_2 e^{-2k_\mu(r_1+r_2)}\frac{1}{r_{12}}$$

$$= -\frac{1}{p_0}\left(\frac{k_\mu^3}{\pi}\right)^2 (4\pi)^2 \int_0^\infty dr_1 r_1^2 e^{-2k_\mu r_1} \int_0^\infty dr_2 r_2^2 e^{-2k_\mu r_2}\frac{1}{r_>}$$

$$(1.47)$$

The radial integral can be evaluated by means of equation (1.42), and the result is:

$$T'_{\nu,\nu} = -\frac{5}{8}\frac{k_\mu}{p_0} = -\frac{1}{\sqrt{2}}\frac{5}{8} \tag{1.48}$$

which is a pure number, independent of p_0. The matrix elements of the interelectron repulsion potential, $T'_{\nu',\nu}$, always prove to be pure numbers, and they are always independent of p_0. The reason for this is that the subsidiary conditions (1.14) and (1.15) require that

$$\frac{k_\mu}{p_0} = \frac{1}{n\sqrt{\frac{1}{n^2}+\frac{1}{n'^2}+...+\frac{1}{n''^2}}} \tag{1.49}$$

Thus the ratios k_μ/p_0 are always a pure numbers; and from equations (1.8), (1.37) and (1.42), it follows that $T'_{\nu',\nu}$ can always be expressed in terms of these ratios. The simple example which we have considered here already allows us to find the energies of the 2-electron isoelectronic series of atoms and ions in the rough approximation where our basis set consists only of a single 2-electron Sturmian basis function - that shown in equation (1.46). In that case, the secular equation reduces to the requirement that

$$p_0 = T^0_{\nu,\nu} + T'_{\nu,\nu} = \sqrt{2}\left(Z - \frac{5}{16}\right) \tag{1.50}$$

where we have made use of equations (1.33) and (1.48). In this rough approximation, the energies of the atoms and ions in the 2-electron isoelectronic series are given by

$$E = -\frac{p_0^2}{2} \approx -\left(Z - \frac{5}{16}\right)^2 \tag{1.51}$$

a well-known result which can be obtained by other methods. Similarly, the ground states of the 3-electron and 4-electron isoelectronic series can be roughly approximated by

$$\phi_\nu(\mathbf{x}) = |\chi_{1s}\chi_{\bar{1}s}\chi_{2s}| \tag{1.52}$$

and

$$\phi_\nu(\mathbf{x}) = |\chi_{1s}\chi_{\bar{1}s}\chi_{2s}\chi_{\bar{2}s}| \tag{1.53}$$

and so on, where the bar indicates a spin-down atomic orbital. Evaluating the Coulomb and exchange integrals by the method discussed above, we obtain ground-state $p_0 \approx T_{\nu,\nu}$ values for the first few N-electron isoelectronic series:

$$p_0 \approx Z\sqrt{\tfrac{2}{1}} - .441942, \qquad N = 2$$

$$p_0 \approx Z\sqrt{\tfrac{2}{1} + \tfrac{1}{4}} - 0.681870, \quad N = 3$$

$$p_0 \approx Z\sqrt{\tfrac{2}{1} + \tfrac{2}{4}} - 0.993588, \quad N = 4$$

$$p_0 \approx Z\sqrt{\tfrac{2}{1} + \tfrac{3}{4}} - 1.40773, \quad N = 5 \tag{1.54}$$

$$p_0 \approx Z\sqrt{\tfrac{2}{1} + \tfrac{4}{4}} - 1.88329, \quad N = 6$$

$$p_0 \approx Z\sqrt{\tfrac{2}{1} + \tfrac{5}{4}} - 2.41491, \quad N = 7$$

The square root multiplying Z in equation (1.54) can be recognized as the square root which appears in equation (1.33), and we can see the way in which this radical reflects the filling of the atomic shells: For all the cases shown in (1.54), the innermost shell contains 2 electrons, while for $N=3,4,5,6$ and 7, there are respectively 1,2,3,4 and 5 electrons in the $n = 2$ shell. Greater accuracy can, of course, be obtained by adding more configurations, but even the crude one-configuration approximation shown in equation (1.54) is in good agreement with Clementi's Hartree-Fock results, as is illustrated in Figure 1. Table 1.1 shows some excited states of the 3-electron isoelectronic series, calculated using a slightly larger basis consisting of 5 generalized Sturmian configurations. As can be seen from the table, good accuracy can also be obtained for

Figure 1.1: Ground-state energies in Hartrees for the 6-electron iso-electronic series: C, N^+, O^{2+}, F^{3+}, Ne^{4+},..etc. as a function of nuclear charge, Z. The smooth curve was calculated from equation (1.54) through the relationship $E = -p_0^2/2$, while the dots show Clementi's Hartree-Fock values [67]. Equation (1.54) becomes highly accurate for large values of Z, while for $Z < N$ the approximation deteriorates and should not be used.

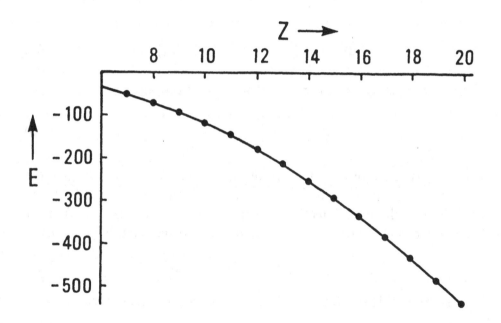

excited states. The accuracy of both the ground-state and excited-states increases when the number of configurations included in the basis set is increased. The evaluation of off-diagonal interconfigurational matrix elements requires the use of the generalized Slater-Condon rules, which are discussed in an appendix.

The many-center problem

Let us end this chapter by looking briefly at the many-center problem, although we shall postpone a detailed treatment of this problem until a later chapter. In the case of molecules, we need to solve the one-electron wave equation

$$\left[-\frac{1}{2}\Delta_j + \frac{1}{2}k_\mu^2 + b_\mu k_\mu v(\mathbf{x}_j) \right] \varphi_\mu(\mathbf{x}_j) = 0 \tag{1.55}$$

where $v(\mathbf{x})$ is the many-center nuclear attraction potential:

$$v(\mathbf{x}_j) = -\sum_a \frac{Z_a}{|\mathbf{x}_j - \mathbf{X}_a|} \tag{1.56}$$

The weighting factor $b_\mu k_\mu$ is introduced into (1.55) in order to give the one-electron orbitals flexibility, so that they may be used to build up a solution to the many-electron equation, (1.1). Equation (1.55) is the many-center analog of (1.11), while the many-center analog of (1.6) is

$$\phi_\nu(\mathbf{x}) = \varphi_\mu(\mathbf{x}_1)\varphi_{\mu'}(\mathbf{x}_2)\varphi_{\mu''}(\mathbf{x}_3)... \tag{1.57}$$

In order that this product should be a solution to (1.1), with

$$V_0(\mathbf{x}) = \sum_{j=1}^{N} v(\mathbf{x}_j) \tag{1.58}$$

we require that the subsidiary conditions

$$k_\mu^2 + k_{\mu'}^2 + k_{\mu''}^2 + ... = -2E \tag{1.59}$$

and

$$b_\mu k_\mu = b_{\mu'} k_{\mu'} = b_{\mu''} k_{\mu''} = ... = \beta_\nu \tag{1.60}$$

be satisfied. Then

$$\frac{1}{2}\Delta\phi_\nu(\mathbf{x})$$

$$= \left[\frac{1}{2}\Delta_1 + \frac{1}{2}\Delta_2 + \ldots\right]\varphi_\mu(\mathbf{x}_1)\varphi_{\mu'}(\mathbf{x}_2)\ldots$$

$$= \left[+\frac{1}{2}k_\mu^2 + b_\mu k_\mu v(\mathbf{x}_1) + \frac{1}{2}k_{\mu'}^2 + b_{\mu'}k_{\mu'}v(\mathbf{x}_2) + \ldots\right]\varphi_\mu(\mathbf{x}_1)\ldots$$

$$= \left[-E + \beta_\nu V_0(\mathbf{x})\right]\phi_\nu(\mathbf{x}) \tag{1.61}$$

Just as in the case of atoms, we of course use Slater determinant basis functions of the form $\phi_\nu(\mathbf{x}) \equiv |\varphi_\mu\varphi_{\mu'}\varphi_{\mu''}\ldots|$ in order that the Pauli principle shall be fulfilled. To solve (1.55), we can expand the orbital $\varphi_\mu(\mathbf{x}_j)$ in terms of hydrogenlike basis functions located on the various atomic centers $\mathbf{X}_a$:

$$\varphi_\mu(\mathbf{x}_j) = \sum_\tau \xi_\tau(\mathbf{x}_j)C_{\tau,\mu} \tag{1.62}$$

where τ stands for the set of indices $\{a, n, l, m, s\}$

$$\xi_\tau(\mathbf{x}_j) \equiv \sqrt{\frac{Z_a}{k_\mu n}}\,\chi_{nlms}(\mathbf{x}_j - \mathbf{X}_a) \tag{1.63}$$

Substituting the expansion (1.62) into the wave equation (1.55) we obtain

$$\sum_\tau \left[-\frac{1}{2}\Delta + \frac{1}{2}k_\mu^2 + b_\mu k_\mu v(\mathbf{x}_j)\right]\xi_\tau(\mathbf{x}_j)C_{\tau,\mu} = 0 \tag{1.64}$$

Next we make use of equation (1.11), which allows us to rewrite (1.64) in the form:

$$\sum_\tau \int d^3x_j\,\xi_{\tau'}^*(\mathbf{x}_j)\left[\frac{n}{|\mathbf{x}_j - \mathbf{X}_a|} + b_\mu v(\mathbf{x}_j)\right]\xi_\tau(\mathbf{x}_j)C_{\tau,\mu} = 0 \tag{1.65}$$

where we have also multiplied from the left by a complex conjugate function from the basis set and integrated over the electron's coordinates. The integrals needed for the solution of the many-center one-electron secular equation can most easily be calculated in momentum space, making use of the properties of hyperspherical harmonics; and therefore we shall discuss these topics in the following chapters.

Table 1.1
Excited ^2S-states of O^{5+} and F^{6+} in Hartrees.

The basis set used consisted of the 5 configurations,
$\phi_{nS} = |\chi_{ns}\chi_{1s}\chi_{\bar{1}s}|$ $n = 2, ..., 6.$
Experimental values are taken from Moore's tables [228].

O^{5+}	$3s\ ^2S$	$4s\ ^2S$	$5s\ ^2S$
calc.	-61.232	-60.267	-59.832
expt.	-61.219	-60.250	-59.813
err.	0.021%	0.028%	0.032%

F^{6+}	$3s\ ^2S$	$4s\ ^2S$	$5s\ ^2S$
calc.	-78.356	-77.057	-76.467
expt.	-78.342	-77.040	-76.450
err.	0.017%	0.022%	0.022%

Table 1.2
One-electron hydrogenlike radial functions, (eq.(1.8)).

$$R_{10}(r) = 2k_1^{3/2}e^{-k_1 r}$$

$$R_{20}(r) = 2k_2^{3/2}(1 - k_2 r)e^{-k_2 r}$$

$$R_{21}(r) = \frac{2k_2^{5/2}}{\sqrt{3}}\, r\, e^{-k_2 r}$$

$$R_{30}(r) = 2k_3^{3/2}\left(1 - 2k_3 r + \frac{2k_3^2 r^2}{3}\right)e^{-k_3 r}$$

$$R_{31}(r) = \frac{(2k_3)^{5/2}}{3}\, r\, \left(1 - \frac{k_3 r}{2}\right)e^{-k_3 r}$$

$$R_{32}(r) = \frac{2^{3/2}k_3^{7/2}}{3\sqrt{5}}\, r^2\, e^{-k_3 r}$$

Exercises

1. Use Table 1.2 to show that the one-electron hydrogenlike Sturmian $\chi_{1,0,0,\frac{1}{2}}(\mathbf{x}_j)$ obeys equations (1.11)-(1.13).

2. Show that if $k_1 = \beta/Z$ and $k_2 = \beta/(2Z)$, then $\chi_{1,0,0,\frac{1}{2}}(\mathbf{x}_j)$ and $\chi_{2,0,0,\frac{1}{2}}(\mathbf{x}_j)$ obey the orthogonality relation

$$\int d\tau_j \, \chi^*_{1,0,0,\frac{1}{2}}(\mathbf{x}_j)\chi_{2,0,0,\frac{1}{2}}(\mathbf{x}_j) = 0$$

3. Show that if $k_1 = k_2 = k_\mu$, then $\chi_{1,0,0,\frac{1}{2}}(\mathbf{x}_j)$ and $\chi_{2,0,0,\frac{1}{2}}(\mathbf{x}_j)$ obey the potential-weighted orthonormality relation

$$\frac{n}{k_\mu} \int d\tau_j \, \chi^*_{nlms}(\mathbf{x}_j)\frac{1}{r}\chi_{n'l'm's'}(\mathbf{x}_j) = \delta_{n'n}\delta_{l'l}\delta_{m'm}\delta_{s's}$$

4. Use equations (1.7) and (1.8) to evaluate the direct-space hydrogenlike orbitals χ_{1s}, χ_{2s}, χ_{2p_j}, χ_{3s} and χ_{3p_j}.

Chapter 2

MOMENTUM-SPACE WAVE FUNCTIONS

Fourier-transformed hydrogenlike orbitals

It is interesting to look at the Fourier transforms of the hydrogenlike orbitals. Dropping the electron index j, we let

$$\chi_{nlm}(\mathbf{x}) = \frac{1}{(2\pi)^{3/2}} \int d^3p \, e^{i\mathbf{p}\cdot\mathbf{x}} \chi_{nlm}^t(\mathbf{p})$$

$$\chi_{nlm}^t(\mathbf{p}) = \frac{1}{(2\pi)^{3/2}} \int d^3x \, e^{-i\mathbf{p}\cdot\mathbf{x}} \chi_{nlm}(\mathbf{x}) \tag{2.1}$$

If we make use of the expansion

$$e^{-i\mathbf{p}\cdot\mathbf{x}} = 4\pi \sum_{l=0}^{\infty} (-i)^l j_l(pr) \sum_{m=-l}^{l} Y_{lm}^*(\theta, \phi) Y_{lm}(\theta_p, \phi_p) \tag{2.2}$$

together with the orthonormality of the spherical harmonics, we obtain:

$$\chi_{nlm}^t(\mathbf{p}) = R_{nl}^t(p) Y_{lm}(\theta_p, \phi_p) \tag{2.3}$$

where

$$R_{nl}^t(p) = (-i)^l \sqrt{\frac{2}{\pi}} \int_0^{\infty} dr \, r^2 j_l(pr) R_{nl}(r) \tag{2.4}$$

23

where $j_l(pr)$ is a spherical Bessel function. The integrals needed for the evaluation of the radial transform are of the form:

$$J_{sl} \equiv \int_0^\infty dr \ r^s e^{-k_\mu r} j_l(pr) \tag{2.5}$$

The most simple of these integrals is

$$J_{10} = \int_0^\infty dr \ r \ e^{-k_\mu r} j_0(pr) = \frac{1}{p} \int_0^\infty dr \ e^{-k_\mu r} \sin(pr) \tag{2.6}$$

which can be evaluated by elementary integration, yielding

$$J_{10} = \frac{1}{p^2 + k_\mu^2} \tag{2.7}$$

By differentiating J_{10} with respect to k_μ we obtain

$$J_{20} = -\frac{\partial}{\partial k_\mu} \frac{1}{p^2 + k_\mu^2} = \frac{2k_\mu}{(p^2 + k_\mu^2)^2} \tag{2.8}$$

$$J_{30} = -\frac{\partial}{\partial k_\mu} \frac{2k_\mu}{(p^2 + k_\mu^2)^2} = \frac{2(3k_\mu^2 - p^2)}{(p^2 + k_\mu^2)^3} \tag{2.9}$$

and so on. The integrals corresponding to other values of l can then be generated using the recursion relations [30, 146]:

$$k_\mu J_{s,l} = p J_{s,l-1} + (s - l - 1) J_{s-1,l} \tag{2.10}$$

and

$$J_{l+1,l} = \frac{2lp}{p^2 + k_\mu^2} J_{l,l-1} \tag{2.11}$$

In general the integrals $J_{l+1,l}$ have the form:

$$J_{l+1,l} = \frac{2^l l! p^l}{(p^2 + k_\mu^2)^{l+1}} \tag{2.12}$$

The first few integrals of this type are shown in Table 2.1; and using them we can derive the Fourier-transformed hydrogenlike orbitals shown in Table 2.2.

Fock's treatment of hydrogenlike atoms

V. Fock [124, 125] introduced the following projection, which maps 3-dimensional p-space onto the surface of a 4-dimensional hypersphere:

$$u_1 = \frac{2k_\mu p_1}{k_\mu^2 + p^2} = \sin\chi \sin\theta_p \cos\phi_p$$

$$u_2 = \frac{2k_\mu p_2}{k_\mu^2 + p^2} = \sin\chi \sin\theta_p \sin\phi_p$$

$$u_3 = \frac{2k_\mu p_3}{k_\mu^2 + p^2} = \sin\chi \cos\theta_p$$

$$u_4 = \frac{k_\mu^2 - p^2}{k_\mu^2 + p^2} = \cos\chi \qquad (2.13)$$

He was then able to show (using a method which we will discuss in a later section) that the Fourier-transformed hydrogenlike orbitals are given by a universal factor,

$$M(p) = \frac{4k_\mu^{5/2}}{(k_\mu^2 + p^2)^2} \qquad (2.14)$$

which is independent of the quantum numbers, multiplied by a 4-dimensional hyperspherical harmonic whose argument is the unit vector defined by equation (2.13):

$$\chi_{n,l,m}^t(\mathbf{p}) = M(p)Y_{n-1,l,m}(\mathbf{u}) \qquad (2.15)$$

The first few 4-dimensional hyperspherical harmonics are shown in Table 2.3 as functions of the unit vector $\mathbf{u} = (u_1, u_2, u_3, u_4)$. These functions are a generalization of the familiar 3-dimensional spherical harmonics; and we will discuss their properties in detail in Chapter 3. The 4-dimensional hyperspherical harmonics are given by [30, 38]

$$Y_{\lambda,l,m}(\mathbf{u}) = \mathcal{N}_{\lambda,l} C_{\lambda-l}^{l+1}(u_4) \sin^l \chi Y_{l,m}(\theta_p, \phi_p) \qquad (2.16)$$

where

$$\mathcal{N}_{\lambda,l} = (-1)^\lambda i^l (2l)!! \sqrt{\frac{2(\lambda+1)(\lambda-l)!}{\pi(\lambda+l+1)!}} \qquad (2.17)$$

and where $C_\lambda^\alpha(u_4)$ is a Gegenbauer polynomial:

$$C_\lambda^\alpha(u_4) = \sum_{t=0}^{[\lambda/2]} \frac{(-1)^t \Gamma(\lambda + \alpha - t)}{t!(\lambda - 2t)!\Gamma(\alpha)} (2u_4)^{\lambda - 2t} \qquad (2.18)$$

The reader may verify, using the hyperspherical harmonics in Table 2.3, that Fock's expression indeed gives the momentum-space hydrogenlike orbitals of Table 2.2 (which we found by laboriously taking the Fourier transforms of the direct-space orbitals). The number of linearly independent 4-dimensional hyperspherical harmonics corresponding to a particular value of the principle quantum number, λ, is given by $(\lambda + 1)^2$; and if we let $n = \lambda + 1$, this corresponds exactly to the degeneracy of the hydrogenlike orbitals. This gives us an explanation of the n^2 degeneracy of the hydrogenlike orbitals, which we would expect to be only $(2l + 1)$-fold degenerate on the basis of the spherical symmetry of the potential. It is striking to see that (apart from the universal factor $M(p)$ and a normalization factor) the hydrogenlike wave functions are represented in momentum space by extremely simple functions of the unit vectors shown in equation (2.13): The $1s$ orbital corresponds to 1; the $2p_1$, $2p_2$, $2p_3$ and $2s$ orbitals correspond to u_1, u_2, u_3 and u_4, while the $3p_1$, $3p_2$ and $3p_3$ orbitals correspond respectively to $u_4 u_1$, $u_4 u_2$ and $u_4 u_3$. We will see in later chapters that momentum-space hydrogenlike orbitals can be used as convenient basis functions for solving many problems in quantum chemistry. Thus hyperspherical harmonics play an important role in momentum-space quantum chemistry; and for this reason, we shall explore some of their properties in Chapter 3. Interestingly, it turns out that there is a d-dimensional generalization for each of the theorems which can be derived for 3-dimensional spherical harmonics.

Table 2.1

$$J_{sl} \equiv \int_0^\infty dr\ r^s e^{-k_\mu r} j_l(pr)$$

l	J_{1l}	J_{2l}	J_{3l}	J_{4l}
0	$\dfrac{1}{p^2 + k_\mu^2}$	$\dfrac{2k_\mu}{(p^2 + k_\mu^2)^2}$	$\dfrac{2(3k_\mu^2 - p^2)}{(p^2 + k_\mu^2)^3}$	$\dfrac{24k_\mu(k_\mu^2 - p^2)}{(p^2 + k_\mu^2)^4}$
1		$\dfrac{2p}{(p^2 + k_\mu^2)^2}$	$\dfrac{8pk_\mu}{(p^2 + k_\mu^2)^3}$	$\dfrac{8p(5k_\mu^2 - p^2)}{(p^2 + k_\mu^2)^4}$
2			$\dfrac{8p^2}{(p^2 + k_\mu^2)^3}$	$\dfrac{48p^2 k_\mu}{(p^2 + k_\mu^2)^4}$
3				$\dfrac{48p^3}{(p^2 + k_\mu^2)^4}$

Table 2.2

Hydrogenlike atomic orbitals and their Fourier transforms

$$\mathbf{t} \equiv k_\mu \mathbf{x}, \qquad t \equiv k_\mu r$$

	$\left(\dfrac{\pi}{k_\mu^3}\right)^{1/2} \chi_{n,l,m}(\mathbf{x})$	$\sqrt{2}\pi\chi_{n,l,m}^t(\mathbf{p})$
$1s$	e^{-t}	$\dfrac{4k_\mu^{5/2}}{(p^2 + k_\mu^2)^2}$
$2s$	$e^{-t}(1 - t)$	$-2\dfrac{4k_\mu^{5/2}}{(p^2 + k_\mu^2)^2}\dfrac{k_\mu^2 - p^2}{p^2 + k_\mu^2}$
$2p_j$	$e^{-t}t_j$	$-2i\dfrac{4k_\mu^{5/2}}{(p^2 + k_\mu^2)^2}\dfrac{2k_\mu p_j}{p^2 + k_\mu^2}$
$3s$	$e^{-t}(1 - 2t + \dfrac{2}{3}t^2)$	$\dfrac{4k_\mu^{5/2}}{(p^2 + k_\mu^2)^2}\left[4\left(\dfrac{k_\mu^2 - p^2}{p^2 + k_\mu^2}\right)^2 - 1\right]$
$3p_j$	$\left(\dfrac{2}{3}\right)^{1/2} e^{-t}(2 - t)t_j$	$i\sqrt{6}\dfrac{4k_\mu^{5/2}}{(p^2 + k_\mu^2)^2}\left(\dfrac{k_\mu^2 - p^2}{p^2 + k_\mu^2}\right)\left(\dfrac{2k_\mu p_j}{p^2 + k_\mu^2}\right)$

Table 2.3
4-dimensional hyperspherical harmonics

λ	l	m	$\sqrt{2}\pi\, Y_{\lambda,l,m}(\mathbf{u})$
0	0	0	1
1	1	1	$i\sqrt{2}(u_1 + iu_2)$
1	1	0	$-i2u_3$
1	1	-1	$-i\sqrt{2}(u_1 - iu_2)$
1	0	0	$-2u_4$

λ	l	m	$\sqrt{2}\pi\, Y_{\lambda,l,m}(\mathbf{u})$
2	2	2	$-\sqrt{3}(u_1 + iu_2)^2$
2	2	1	$2\sqrt{3}u_3(u_1 + iu_2)$
2	2	0	$-\sqrt{2}(2u_3^2 - u_1^2 - u_2^2)$
2	2	-1	$-2\sqrt{3}u_3(u_1 - iu_2)$
2	2	-2	$-\sqrt{3}(u_1 - iu_2)^2$
2	1	1	$-i2\sqrt{3}\, u_4(u_1 + iu_2)$
2	1	0	$2i\sqrt{6}\, u_4 u_3$
2	1	-1	$2i\sqrt{3}\, u_4(u_1 - iu_2)$
2	0	0	$4u_4^2 - 1$

λ	l	m	$\sqrt{2}\pi\, Y_{\lambda,l,m}(\mathbf{u})$
3	3	3	$-2i(u_1 + iu_2)^3$
3	3	2	$i\sqrt{24}u_3(u_1 + iu_2)^2$
3	3	1	$2i\sqrt{\dfrac{3}{5}}(u_1 + iu_2)(u_1^2 + u_2^2 - 4u_3^2)$
3	3	0	$-\dfrac{4i}{\sqrt{5}}u_3(3u_1^2 + 3u_2^2 - 2u_3^2)$
3	2	2	$\sqrt{24}u_4(u_1 + iu_2)^2$
3	2	1	$-\sqrt{96}(u_1 + iu_2)u_3u_4$
3	2	0	$-4u_4(u_1^2 + u_2^2 - 2u_3^2)$
3	1	1	$\sqrt{\dfrac{8}{5}}(u_1 + iu_2)(6u_4^2 - 1)$
3	1	0	$i\dfrac{4}{\sqrt{5}}u_3(6u_4^2 - 1)$
3	0	0	$4u_4(1 - 2u_4^2)$

Table 2.4
Alternative 4-dimensional hyperspherical harmonics

τ	λ	l	$\sqrt{2}\pi\, Y_\tau(\mathbf{u})$
1s	0	0	1
$2p_1$	1	1	$-2iu_1$
$2p_2$	1	1	$-2iu_2$
$2p_3$	1	1	$-2iu_3$
2s	1	0	$-2u_4$

$3d_{z^2}$	2	2	$-\sqrt{2}(2u_3^2 - u_1^2 - u_2^2)$
$3d_{x^2-y^2}$	2	2	$-\sqrt{6}(u_1^2 - u_2^2)$
$3d_{xy}$	2	2	$-2\sqrt{6}\, u_1u_2$
$3d_{yz}$	2	2	$-2\sqrt{6}\, u_2u_3$
$3d_{zx}$	2	2	$-2\sqrt{6}\, u_3u_1$
$3p_1$	2	1	$2i\sqrt{6}\, u_4u_1$
$3p_2$	2	1	$2i\sqrt{6}\, u_4u_2$
$3p_3$	2	1	$2i\sqrt{6}\, u_4u_3$
3s	2	0	$4u_4^2 - 1$

Exercises

1. Calculate the integral in equation (2.6) and show that it yields the result shown (2.7).

2. Starting with J_{10}, calculate the integrals J_{20} and J_{30} by differentiating with respect to k_μ, as shown in equations (2.8) and (2.9).

3. Starting with J_{10}, use the recursion relation of equation (2.11) to generate J_{21}, J_{31} and J_{32}.

4. Use the integrals J_{sl} in Table 2.1 to evaluate the Fourier transforms of the direct-space hydrogenlike orbitals χ_{1s}, χ_{2s} and χ_{2p_j}. Show that the transforms correspond to the solutions of Fock, equation (2.15).

Chapter 3

HYPERSPHERICAL HARMONICS

Harmonic polynomials

In the previous chapter, we saw that when 3-dimensional momentum space is mapped onto the surface of a 4-dimensional hypersphere by Fock's projection, the Fourier-transformed hydrogenlike atomic orbitals are represented by 4-dimensional hyperspherical harmonics, multiplied by a universal factor, $M(p)$, which is independent of the quantum numbers. We mentioned also that our aim will be to use the momentum-space hydrogenlike functions as basis sets for constructing atomic and molecular orbitals, as well as crystal orbitals. Thus, hyperspherical harmonics play an important role in momentum-space quantum chemistry; and for this reason, we shall devote the present chapter to discussing some of their properties.

The simplest approach to hyperspherical harmonics is through the theory of harmonic polynomials [30, 38, 296, 300]: If we consider a d-dimensional space, with Cartesian coordinates $x_1, x_2, ..., x_d$, then a *harmonic* polynomial h_λ is, by definition, a homogeneous polynomial which satisfies the generalized Laplace equation:

$$\Delta h_\lambda = 0 \tag{3.1}$$

where Δ is defined by equation (1.2). A *homogeneous* polynomial f_n of order n is defined as a polynomial which is built up from a linear

33

superposition of monomial terms of the form

$$m_n \equiv \prod_{j=1}^{d} x_j^{n_j} \tag{3.2}$$

where

$$n_1 + n_2 + ... + n_d = n \tag{3.3}$$

Thus, for example, $2x_1^2 + x_1 x_2$ is a homogeneous polynomial of order 2, but it is not harmonic; whereas $x_1^2 - x_2^2$ is a harmonic polynomial of order 2. Harmonic polynomials in a d-dimensional space are related to hyperspherical harmonics by

$$h_\lambda = r^\lambda Y_\lambda \tag{3.4}$$

where r is the *hyperradius*, defined by

$$r^2 \equiv \sum_{j=1}^{d} x_j^2 \tag{3.5}$$

Thus, for example, apart from a normalization factor,

$$Y_{2,x_1^2-x_2^2} \sim \frac{x_1^2 - x_2^2}{r^2} \tag{3.6}$$

is a hyperspherical harmonic with principal quantum number $\lambda = 2$ in our d-dimensional space.

Any homogeneous polynomial f_n can be decomposed into a series of harmonic polynomials multiplied by powers of the hyperradius:

$$f_n = h_n + r^2 h_{n-2} + r^4 h_{n-4} + ... \tag{3.7}$$

In order to carry out this decomposition, we first notice that if we differentiate the monomial m_n of equation (3.2) with respect to one of the coordinates, we obtain:

$$\frac{\partial m_n}{\partial x_j} = n_j x_j^{-1} m_n \tag{3.8}$$

Multiplying both sides of (3.8) by x_j and summing over j, we have:

$$\sum_{j=1}^{d} x_j \frac{\partial m_n}{\partial x_j} = n m_n \qquad (3.9)$$

where $n = n_1 + n_2 + ... + n_d$. Since this result holds for each of the monomials in a homogeneous polynomial of order n, it must hold for the entire polynomial. Thus

$$\sum_{j=1}^{d} x_j \frac{\partial f_n}{\partial x_j} = n f_n \qquad (3.10)$$

where f_n represents any homogeneous polynomial of order n. From (3.10) it follows that if r^β is the hyperradius raised to some power β, then

$$\Delta \left(r^\beta f_\alpha \right) = \beta(\beta + d + 2\alpha - 2) r^{\beta-2} f_\alpha + r^\beta \Delta f_\alpha \qquad (3.11)$$

If h_α is a harmonic polynomial of order α, it follows from (3.11) that

$$\Delta \left(r^\beta h_\alpha \right) = \beta(\beta + d + 2\alpha - 2) r^{\beta-2} h_\alpha \qquad (3.12)$$

since

$$\Delta h_\alpha = 0 \qquad (3.13)$$

Equation (3.12) is just what we need in order to decompose a homogeneous polynomial f_n into its harmonic components. If we apply the generalized Laplacian operator Δ repeatedly to both sides of (3.7), we obtain:

$$\begin{aligned}
\Delta f_n &= 2(d + 2n - 4) h_{n-2} + 4(d + 2n - 6) r^2 h_{n-4} + ... \\
\Delta^2 f_n &= 8(d + 2n - 6)(d + 2n - 8) h_{n-4} + ...
\end{aligned} \qquad (3.14)$$

and in general,

$$\Delta^\nu f_n = \sum_{k=\nu}^{\left[\frac{1}{2}n\right]} \frac{(2k)!!}{(2k - 2\nu)!!} \frac{(d + 2n - 2k - 2)!!}{(d + 2n - 2k - 2\nu - 2)!!} r^{2k-2\nu} h_{n-2k} \qquad (3.15)$$

where $[n/2]$ means "the largest integer in $[n/2]$", and $n!! \equiv n(n-2)(n-4)...2$ when n is even or $n!! \equiv n(n-2)(n-4)...1$ when n is odd, and where $0!! \equiv 1$. This gives us a set of simultaneous equations which can be solved to yield expressions for the harmonic polynomials in the series (3.7). For example, the relationship

$$\Delta^{\frac{1}{2}n} f_n = \frac{n!!(d+n-2)!!}{(d-2)!!} h_0 \tag{3.16}$$

can be solved to give an expression for the constant h_0:

$$h_0 = \frac{(d-2)!!}{n!!(d+n-2)!!} \Delta^{\frac{1}{2}n} f_n \tag{3.17}$$

Solving (3.15) for the first few values of n, we obtain:

$$
\begin{aligned}
f_2 &= h_2 + r^2 h_0 \\
h_0 &= \frac{1}{2d} \Delta f_2 \\
h_2 &= f_2 - \frac{r^2}{2d} \Delta f_2 \\[2mm]
f_3 &= h_3 + r^2 h_1 \\
h_1 &= \frac{1}{2(d+2)} \Delta f_3 \\
h_3 &= f_3 - \frac{r^2}{2(d+2)} \Delta f_3
\end{aligned}
$$

$$\tag{3.18}$$

and so on. Thus, for example, suppose that $f_2 = 2x_1^2 + x_1 x_2$. If we wish to decompose this homogeneous polynomial into its harmonic components, we first note that $\Delta f_2 = 4$. Then from (3.18) we have:

$$
\begin{aligned}
f_2 &= h_2 + r^2 h_0 = 2x_1^2 + x_1 x_2 \\
h_0 &= \frac{1}{2d} \Delta f_2 = \frac{2}{d} \\
h_2 &= f_2 - \frac{r^2}{2d} \Delta f_2 = 2x_1^2 + x_1 x_2 - \frac{2r^2}{d}
\end{aligned}
\tag{3.19}
$$

From these examples, and also from (3.15), we can see that the harmonic polynomial of highest order in the decomposition of f_n has the form:

$$h_n = f_n + a_2 r^2 \Delta f_n + a_4 r^4 \Delta^2 f_n + a_6 r^6 \Delta^3 f_n + \dots \tag{3.20}$$

If we apply the generalized Laplacian operator, Δ, to both sides of (3.20), making use of the fact that $\Delta h_n = 0$, we obtain

$$0 = \Delta f_n + a_2 \Delta(r^2 \Delta f_n) + a_4 \Delta(r^4 \Delta^2 f_n) + \dots \tag{3.21}$$

We now make use of (3.12) and collect terms in $\Delta^\nu f_n$. In order for (3.21) to hold for all values of $\mathbf{x}$, the sum of the terms corresponding to each value of ν must vanish. This condition gives us a series of equations for the constants a_j:

$$
\begin{aligned}
1 + 2(d + 2n - 4)a_2 &= 0 \\
a_2 + 4(d + 2n - 6)a_4 &= 0 \\
a_4 + 6(d + 2n - 8)a_6 &= 0 \\
\vdots \quad &= \quad \vdots
\end{aligned}
\tag{3.22}
$$

which can be solved to yield

$$
\begin{aligned}
h_n &= f_n - \frac{r^2}{2(d + 2n - 4)}\Delta f_n + \frac{r^4}{8(d + 2n - 4)(d + 2n - 6)}\Delta^2 f_n + \dots \\
&= \sum_{j=0}^{[n/2]} \frac{(-1)^j(d + 2n - 2j - 4)!!}{(2j)!!(d + 2n - 4)!!} r^{2j} \Delta^j f_n
\end{aligned}
\tag{3.23}
$$

This gives us an explicit expression for the harmonic polynomial of highest order in the decomposition of f_n. In a similar way, one can show that the harmonic polynomials of lower order in the decomposition are given by

$$
\begin{aligned}
h_{n-2\nu} &= \frac{(d + 2n - 4\nu - 2)!!}{(2\nu)!!(d + 2n - 2\nu - 2)!!} \\
&\times \sum_{j=0}^{[\frac{n}{2} - \nu]} \frac{(-1)^j(d + 2n - 4\nu - 2j - 4)!!}{(2j)!!(d + 2n - 4\nu - 4)!!} r^{2j} \Delta^{j+\nu} f_n
\end{aligned}
\tag{3.24}
$$

Table 2.1 shows the decomposition of polynomials into their harmonic components for the first few values of n.

Grand angular momentum

The grand angular momentum operator Λ^2 in a d-dimensional space is defined by the relation:

$$\Lambda^2 \equiv -\sum_{i>j}^{d} \left(x_i \frac{\partial}{\partial x_j} - x_j \frac{\partial}{\partial x_i} \right)^2 \qquad (3.25)$$

Alternatively, we can express Λ^2 in the form:

$$\Lambda^2 = -r^2 \Delta + \sum_{i,j=1}^{d} x_i x_j \frac{\partial^2}{\partial x_i \partial x_j} + (d-1) \sum_{i=1}^{d} x_i \frac{\partial}{\partial x_i} \qquad (3.26)$$

where Δ is the generalized Laplace operator:

$$\Delta = \sum_{i=1}^{d} \frac{\partial^2}{\partial x_i^2} \qquad (3.27)$$

In the previous section, we proved that for any homogeneous polynomial f_n of order n,

$$\sum_{i=1}^{d} x_i \frac{\partial f_n}{\partial x_i} = n f_n \qquad (3.28)$$

One can show in a similar way that

$$\sum_{i,j=1}^{d} x_i x_j \frac{\partial^2 f_n}{\partial x_i \partial x_j} = n(n-1) f_n \qquad (3.29)$$

From equations (3.26)-(3.29), it follows that if we apply the grand angular momentum operator to any homogeneous polynomial, we obtain:

$$\Lambda^2 f_n = -r^2 \Delta f_n + n(n+d-2) f_n \qquad (3.30)$$

Furthermore, if Λ^2 is applied to a harmonic polynomial h_λ, (3.30) reduces to:

$$\Lambda^2 h_\lambda = \lambda(\lambda + d - 2) h_\lambda \qquad (3.31)$$

Thus any harmonic polynomial of order λ in a d-dimensional space is an eigenfunction of Λ^2 corresponding to the eigenvalue $\lambda(\lambda + d - 2)$;

and the decomposition of a homogeneous polynomial f_n into harmonic polynomials is, in fact, a resolution of f_n into eigenfunctions of grand angular momentum. Notice that in the case where $d = 3$, $\lambda(\lambda + d - 2)$ reduces to the familiar orbital angular momentum eigenvalue, $l(l + 1)$, provided that we make the identification, $\lambda \to l$.

From the definition of the hyperradius, (3.5), it follows that

$$r\frac{\partial}{\partial r} = \sum_{i=1}^{d} x_i \frac{\partial}{\partial x_i} \tag{3.32}$$

and

$$r^2\frac{\partial^2}{\partial r^2} = \sum_{i,j=1}^{d} x_i x_j \frac{\partial^2}{\partial x_i \partial x_j} \tag{3.33}$$

These two relations can be combined with equation (3.26) to yield an expression for the generalized Laplacian operator in terms an operator involving the hyperradius and Λ^2:

$$\Delta = \sum_{j=1}^{d} \frac{\partial^2}{\partial x_j^2} = \frac{1}{r^{d-1}}\frac{\partial}{\partial r}r^{d-1}\frac{\partial}{\partial r} - \frac{\Lambda^2}{r^2} \tag{3.34}$$

For example, when $d = 3$, this reduces to the familiar relationship:

$$\Delta = \frac{1}{r^2}\frac{\partial}{\partial r}r^2\frac{\partial}{\partial r} - \frac{L^2}{r^2} \tag{3.35}$$

Angular integrations

In a d-dimensional space, the volume element is given by

$$dx_1 dx_2...dx_d = r^{d-1}dr\, d\Omega \tag{3.36}$$

where $d\Omega$ is the generalized solid angle element. The total solid angle can be evaluated in the following way: Consider the integral of e^{-r^2}, taken over all space.

$$\int_0^\infty dr\, r^{d-1}e^{-r^2} \int d\Omega = \prod_{j=1}^{d} \int_{-\infty}^{\infty} dx_j e^{-x_j^2} \tag{3.37}$$

On the left-hand side of equation (3.37), we express this integral in terms of the hyperradius and hyperangles, while on the right-hand side, we express it in terms of the individual Cartesian coordinates. The integral over the hyperradius can be evaluated in terms of the gamma function:

$$\int_0^\infty dr\, r^{d-1} e^{-r^2} = \frac{\Gamma(d/2)}{2} \tag{3.38}$$

and similarly, for each of the integrals on the right, we have

$$\int_{-\infty}^\infty dx_j e^{-x_j^2} = \Gamma(1/2) = \pi^{\frac{1}{2}} \tag{3.39}$$

Combining equations (3.37)-(3.39), we obtain an expression for the total solid angle:

$$\int d\Omega = \frac{2\pi^{\frac{d}{2}}}{\Gamma\left(\frac{d}{2}\right)} \equiv I(0) \tag{3.40}$$

For example, when $d = 3$, this expression yields the familiar result:

$$I(0) = \frac{2\pi^{\frac{3}{2}}}{\Gamma\left(\frac{3}{2}\right)} = 4\pi \tag{3.41}$$

while when $d = 4$ it gives

$$I(0) = \frac{2\pi^2}{\Gamma(2)} = 2\pi^2 \tag{3.42}$$

It is interesting to consider the integral of the product of two harmonic polynomials h_λ and $h_{\lambda'}$, taken over the generalized solid angle. If $\lambda \neq \lambda'$, i.e. if the orders of the two harmonic polynomials are different, then (as we saw in the previous section) they are eigenfunctions of the grand angular momentum operator Λ^2 belonging to different eigenvalues. In that case, it follows from the Hermiticity of the operator Λ^2 with respect to integration over $d\Omega$ that

$$\int d\Omega\, h_\lambda^* h_{\lambda'} = 0 \quad \text{if } \lambda \neq \lambda' \tag{3.43}$$

For the case where $\lambda' = 0$, (3.43) reduces to

$$\int d\Omega\, h_\lambda = 0 \quad \text{if } \lambda \neq 0 \tag{3.44}$$

since h_0 is just a constant. Equations (3.44) and (3.7) allow us to calculate the integral of any homogeneous polynomial f_n over the generalized solid angle, since we can write

$$\int d\Omega\, f_n = \int d\Omega\, (h_n + r^2 h_{n-2} + \dots + r^n h_0) \tag{3.45}$$

Because of (3.44), all of the terms in this integral vanish except the last one, and we obtain [27, 39]:

$$\int d\Omega\, f_n = I(0) r^n h_0 \tag{3.46}$$

or, from (3.17) and (3.40),

$$\int d\Omega\, f_n = \frac{2\pi^{d/2} r^n (d-2)!!}{\Gamma(d/2) n!! (d+n-2)!!} \Delta^{\frac{1}{2}n} f_n \tag{3.47}$$

This powerful formula allows us to integrate by differentiation! If

$$f_n = m_n = \prod_{j=1}^{d} x_j^{n_j} \qquad n_1 + n_2 + \dots + n_d = n \tag{3.48}$$

then, if all the n_j's are even, we obtain

$$I(\mathbf{n}) \equiv r^{-n} \int d\Omega \prod_{j=1}^{d} x_j^{n_j} = \frac{2\pi^{d/2}(d-2)!!}{\Gamma(d/2)(d+n-2)!!} \prod_{j=1}^{d} (n_j - 1)!! \tag{3.49}$$

while if any of the n_j's are uneven, the integral vanishes. It is possible to generalize equation (3.47) and to derive an expression for the angular integral of an arbitrary function, provided that it is possible to expand the function about the origin in terms of polynomials in the Cartesian coordinates $x_1, x_2, \dots, x_d$ [43]. If such an expansion is possible, then we can write

$$F(\mathbf{x}) = \sum_{n=0}^{\infty} f_n(\mathbf{x}) \tag{3.50}$$

where the functions $f_n(\mathbf{x})$ are homogeneous polynomials. Then from (3.47) we have:

$$\int d\Omega F(\mathbf{x}) = \frac{(d-2)!! 2\pi^{d/2}}{\Gamma\left(\frac{d}{2}\right)} \sum_{n=0,2,\dots}^{\infty} \frac{r^n}{n!!(n+d-2)!!} \Delta^{n/2} f_n(\mathbf{x}) \tag{3.51}$$

where the odd terms have been omitted because they cannot contribute to the angular integral. From (3.50) it follows that

$$\left[\Delta^{n/2}F(\mathbf{x})\right]_{\mathbf{x}=0} = \Delta^{n/2}f_n(\mathbf{x}) \qquad (3.52)$$

since the operation of setting $\mathbf{x} = 0$ eliminates all parts of a polynomial in $x_1, x_2, ..., x_d$ except the constant term. Finally, if we let $n = 2\nu$, making use of (3.52), we can see that (3.51) will take on the form

$$\int d\Omega F(\mathbf{x}) = \frac{(d-2)!!2\pi^{d/2}}{\Gamma\left(\frac{d}{2}\right)} \sum_{\nu=0}^{\infty} \frac{r^{2\nu}}{(2\nu)!!(d+2\nu-2)!!} \left[\Delta^{\nu}F(\mathbf{x})\right]_{\mathbf{x}=0} \qquad (3.53)$$

When $d = 3$, this formula reduces to [43]

$$\int d\Omega F(\mathbf{x}) = 4\pi \sum_{\nu=0}^{\infty} \frac{r^{2\nu}}{(2\nu+1)!} \left[\Delta^{\nu}F(\mathbf{x})\right]_{\mathbf{x}=0} \qquad (3.54)$$

To illustrate equation (3.53), we can consider the case where $F(\mathbf{x})$ is a d-dimensional plane wave:

$$F(\mathbf{x}) = e^{i\mathbf{k}\cdot\mathbf{x}} = e^{i(k_1 x_1 + k_2 x_2 + ... + k_d x_d)} \qquad (3.55)$$

Then

$$\left[\Delta^{\nu}e^{i\mathbf{k}\cdot\mathbf{x}}\right]_{\mathbf{x}=0} = (-1)^{\nu}k^{2\nu} \qquad (3.56)$$

so that (3.53) yields

$$\int d\Omega e^{i\mathbf{k}\cdot\mathbf{x}} = \frac{(d-2)!!2\pi^{d/2}}{\Gamma\left(\frac{d}{2}\right)} \sum_{\nu=0}^{\infty} \frac{(-1)^{\nu}(kr)^{2\nu}}{(2\nu)!!(d+2\nu-2)!!} \qquad (3.57)$$

When $d = 3$, this reduces to

$$\int d\Omega e^{i\mathbf{k}\cdot\mathbf{x}} = 4\pi \sum_{\nu=0}^{\infty} \frac{(-1)^{\nu}(kr)^{2\nu}}{(2\nu+1)!} = 4\pi j_0(kr) \qquad (3.58)$$

where $j_0(kr)$ is a spherical Bessel function of order zero. Notice that in this example, we did not actually expand the function $F(\mathbf{x})$ in terms of polynomials in the coordinates $x_1, x_2, ..., x_d$, although it was necessary to assume that such an expansion could be made.

Hyperspherical harmonics

Suppose that we have several linearly independent harmonic polynomials of order λ. We will then need additional labels, besides λ, to distinguish between them; and we can call this set of indices $\{\mu\}$. All of the harmonic polynomials in such a set will be eigenfunctions of the grand angular momentum operator Λ^2 corresponding to the same eigenvalue:

$$\Lambda^2 h_{\lambda\{\mu\}} = \lambda(\lambda + d - 1)h_{\lambda\{\mu\}} \tag{3.59}$$

We can choose the linearly independent harmonic polynomials in such a way that they will satisfy orthonormality relations of the form:

$$\int d\Omega\, h^*_{\lambda'\{\mu'\}}h_{\lambda\{\mu\}} = \delta_{\lambda'\lambda}\delta_{\{\mu'\}\{\mu\}}r^{\lambda+\lambda'} \tag{3.60}$$

There will then be a set of hyperspherical harmonics related to the harmonic polynomials by equation (3.4):

$$h_{\lambda\{\mu\}} = r^\lambda Y_{\lambda\{\mu\}} \tag{3.61}$$

and the hyperspherical harmonics will obey the orthonormality relations

$$\int d\Omega\, Y^*_{\lambda'\{\mu'\}}Y_{\lambda\{\mu\}} = \delta_{\lambda'\lambda}\delta_{\{\mu'\}\{\mu\}} \tag{3.62}$$

In the case of the 4-dimensional hyperspherical harmonics shown in Table 1.1, the set of indices $\{\mu\}$ corresponds to $\{l, m\}$. These indices are labels for the irreducible representations of a chain of subgroups of the 4-dimensional rotation group $SO(4)$.

$$SO(4) \supset SO(3) \supset SO(2) \tag{3.63}$$

The standard chain of subgroups used in constructing orthonormal sets of hyperspherical harmonics in pure mathematics is [296]

$$SO(d) \supset SO(d-1) \supset SO(d-2).... \supset SO(2) \tag{3.64}$$

but this choice is not necessarily the most convenient one in physical applications. In a later chapter we will discuss the chains of subgroups of $SO(d)$ which are appropriate for constructing orthonormal sets of hyperspherical harmonics adapted to the symmetry of various quantum mechanical problems.

Gegenbauer polynomials

Each of the familiar theorems which one can derive for 3-dimensional spherical harmonics has a d-dimensional generalization. In the theory of hyperspherical harmonics, Legendre polynomials are replaced by Gegenbauer polynomials, which are defined by the generating function:

$$\frac{1}{|\mathbf{x} - \mathbf{x}'|^{d-2}} = \frac{1}{r_>^{d-2}(1 + \epsilon^2 - 2\epsilon\hat{\mathbf{x}} \cdot \hat{\mathbf{x}}')^\alpha} = \frac{1}{r_>^{d-2}} \sum_{\lambda=0}^{\infty} \left(\frac{r_<}{r_>}\right)^\lambda C_\lambda^\alpha(\hat{\mathbf{x}} \cdot \hat{\mathbf{x}}')$$

(3.65)

where

$$\epsilon \equiv \frac{r_<}{r_>} \tag{3.66}$$

$$\alpha \equiv \frac{d-2}{2} \tag{3.67}$$

and

$$\hat{\mathbf{x}} \equiv \frac{\mathbf{x}}{r} = \frac{1}{r}(x_1, x_2, ..., x_d)$$
$$\hat{\mathbf{x}}' \equiv \frac{\mathbf{x}'}{r'} = \frac{1}{r'}(x_1', x_2', ..., x_d') \tag{3.68}$$

The polynomials defined by this generating function are given by [153]:

$$C_\lambda^\alpha(\hat{\mathbf{x}} \cdot \hat{\mathbf{x}}') = \sum_{t=0}^{[\lambda/2]} \frac{(-1)^t \Gamma(\alpha + \lambda - t)(2\hat{\mathbf{x}} \cdot \hat{\mathbf{x}}')^{\lambda-2t}}{t!(\lambda - 2t)!\Gamma(\alpha)}$$

$$= \frac{1}{(d-4)!!} \sum_{t=0}^{[\lambda/2]} \frac{(-1)^t(d + 2\lambda - 2t - 4)!!}{(2t)!!(\lambda - 2t)!}(\hat{\mathbf{x}} \cdot \hat{\mathbf{x}}')^{\lambda-2t}$$

(3.69)

Equation (3.65) is the d-dimensional generalization of the familiar expansion of the Green's function of the Laplacian operator in terms of Legendre polynomials:

$$\frac{1}{|\mathbf{x} - \mathbf{x}'|} = \frac{1}{r_>(1 + \epsilon^2 - 2\epsilon\hat{\mathbf{x}} \cdot \hat{\mathbf{x}}')^{1/2}} = \frac{1}{r_>} \sum_{l=0}^{\infty} \left(\frac{r_<}{r_>}\right)^l P_l(\hat{\mathbf{x}} \cdot \hat{\mathbf{x}}') \quad (3.70)$$

and in fact, the function $|\mathbf{x} - \mathbf{x}'|^{2-d}$ is the Green's function of the generalized Laplacian operator, (3.27). When $d = 3$ and $\alpha = 1/2$, the

Gegenbauer polynomials reduce to Legendre polynomials. Table 3.2 shows the first few Gegenbauer polynomials for the case where d and α are general, and in the special case where $d = 4$ and $\alpha = 1$.

The familiar 3-dimensional spherical harmonics obey a sum rule involving Legendre polynomials:

$$\frac{4\pi}{2l+1} \sum_m Y_{lm}^*(\hat{\mathbf{x}}')Y_{lm}(\hat{\mathbf{x}}) = P_l(\hat{\mathbf{x}} \cdot \hat{\mathbf{x}}') \tag{3.71}$$

Similarly, one can show that hyperspherical harmonics obey a sum rule involving Gegenbauer polynomials:

$$\frac{(d-2)I(0)}{2\lambda + d - 2} \sum_{\{\mu\}} Y_{\lambda\{\mu\}}^*(\hat{\mathbf{x}}')Y_{\lambda\{\mu\}}(\hat{\mathbf{x}}) = C_\lambda^\alpha(\hat{\mathbf{x}} \cdot \hat{\mathbf{x}}') \tag{3.72}$$

The sum rule for hyperspherical harmonics can be established in the following way: If we let $\mathbf{x} - \mathbf{x}' \equiv \mathbf{x}_-$, and note that $\Delta = \Delta_-$, then from (3.34) we have

$$\Delta \frac{1}{|\mathbf{x} - \mathbf{x}'|^{d-2}} = \left(\frac{1}{r_-^{d-1}} \frac{\partial}{\partial r_-} r_-^{d-1} \frac{\partial}{\partial r_-} - \frac{\Lambda^2}{r_-^2}\right) r_-^{2-d} = 0 \quad r_- \neq 0 \tag{3.73}$$

We now replace the Green's function of the generalized Laplacian operator by its expansion in terms of Gegenbauer polynomials, (3.65). Equation (3.73) then becomes:

$$\Delta \frac{1}{|\mathbf{x} - \mathbf{x}|^{d-2}} = \sum_{\lambda=0}^\infty \frac{1}{r'^{\lambda+d-2}} \Delta \left[r^\lambda C_\lambda^\alpha(\hat{\mathbf{x}} \cdot \hat{\mathbf{x}}')\right] = 0 \tag{3.74}$$

For the expansion in equation (3.74) to hold for all values of the coordinates, each term must vanish separately, and thus

$$\left(\frac{1}{r^{d-1}} \frac{\partial}{\partial r} r^{d-1} \frac{\partial}{\partial r} - \frac{\Lambda^2}{r^2}\right) r^\lambda C_\lambda^\alpha(\hat{\mathbf{x}} \cdot \hat{\mathbf{x}}') = 0 \tag{3.75}$$

The radial operator in Δ, acting on r^λ, gives

$$\frac{1}{r^{d-1}} \frac{\partial}{\partial r} r^{d-1} \frac{\partial}{\partial r} r^\lambda = \lambda(\lambda + d - 2)r^{\lambda-2} \tag{3.76}$$

and therefore

$$\left[\Lambda^2 - \lambda(\lambda + d - 2)\right] C_\lambda^\alpha(\hat{\mathbf{x}} \cdot \hat{\mathbf{x}}') = 0 \tag{3.77}$$

In other words, $C_\lambda^\alpha(\hat{\mathbf{x}} \cdot \hat{\mathbf{x}}')$ is an eigenfunction of the grand angular momentum operator corresponding to the eigenvalue, $\lambda(\lambda + d - 2)$. However, we know that the Hilbert space of all such functions is spanned by the hyperspherical harmonics belonging to this eigenvalue. Therefore we must be able to express $C_\lambda^\alpha(\hat{\mathbf{x}} \cdot \hat{\mathbf{x}}')$ as a linear combination of this set of hyperspherical harmonics:

$$C_\lambda^\alpha(\hat{\mathbf{x}} \cdot \hat{\mathbf{x}}') = \sum_{\{\mu\}} a_{\lambda\{\mu\}}(\hat{\mathbf{x}}') Y_{\lambda\{\mu\}}(\hat{\mathbf{x}}) \tag{3.78}$$

From the invariance of $C_\lambda^\alpha(\hat{\mathbf{x}} \cdot \hat{\mathbf{x}}')$ under coordinate rotations, it follows that

$$a_{\lambda\{\mu\}}(\hat{\mathbf{x}}') = K_\lambda Y_{\lambda\{\mu\}}^*(\hat{\mathbf{x}}') \tag{3.79}$$

where K_λ is some constant. Thus,

$$C_\lambda^\alpha(\hat{\mathbf{x}} \cdot \hat{\mathbf{x}}') = K_\lambda \sum_{\{\mu\}} Y_{\lambda\{\mu\}}^*(\hat{\mathbf{x}}') Y_{\lambda\{\mu\}}(\hat{\mathbf{x}}) \tag{3.80}$$

If $F(\hat{\mathbf{x}})$ is a function of the hyperangles in a d-dimensional space, then it follows from (3.71) that

$$O_\lambda[F(\hat{\mathbf{x}})] \equiv \frac{1}{K_\lambda} \int d\Omega' C_\lambda^\alpha(\hat{\mathbf{x}} \cdot \hat{\mathbf{x}}') F(\hat{\mathbf{x}}') = \sum_{\{\mu\}} Y_{\lambda\{\mu\}}(\hat{\mathbf{x}}) \int d\Omega' Y_{\lambda\{\mu\}}^*(\hat{\mathbf{x}}') F(\hat{\mathbf{x}}') \tag{3.81}$$

will be a projection of $F(\hat{\mathbf{x}})$ onto the part of Hilbert space spanned by eigenfunctions of Λ^2 corresponding to the eigenvalue $\lambda(\lambda + d - 2)$. Of course, if we perform the projection on a function which is entirely within this subspace, the function will be unchanged; and this condition can be used to show that the constant K_λ is given by:

$$K_\lambda = \frac{(d - 2)I(0)}{2\lambda + d - 2} \tag{3.82}$$

where $I(0)$ is the total generalized solid angle.

The sum rule tells us immediately how many linearly independent hyperspherical harmonics there are for each value of λ: For the case where $\hat{\mathbf{x}}' = \hat{\mathbf{x}}$, the expansion shown in equation (3.65) yields

$$C_\lambda^\alpha(1) = \frac{(\lambda + d - 3)!}{\lambda!(d - 3)!} \tag{3.83}$$

and (3.80) reduces to:

$$C_\lambda^\alpha(1) = K_\lambda \sum_{\{\mu\}} Y_{\lambda\{\mu\}}^*(\hat{\mathbf{x}}) Y_{\lambda\{\mu\}}(\hat{\mathbf{x}}) \tag{3.84}$$

Integrating (3.84) over the generalized solid angle, and making use of the normalization of the hyperspherical harmonics, we obtain:

$$C_\lambda^\alpha(1) \int d\Omega = K_\lambda \sum_{\{\mu\}} 1 = K_\lambda \omega \tag{3.85}$$

where

$$\omega = \frac{1}{K_\lambda} C_\lambda^\alpha(1) I(0) \tag{3.86}$$

is the degeneracy of the hyperspherical harmonics, i.e., the number of harmonics corresponding to the principal quantum number λ. Combining these relationships, we obtain an explicit expression for the degeneracy:

$$\omega = \frac{(d + 2\lambda - 2)(\lambda + d - 3)!}{\lambda!(d - 2)!} \tag{3.87}$$

For example, when $d = 3$,

$$\omega = 2\lambda + 1 \tag{3.88}$$

while when $d = 4$,

$$\omega = (\lambda + 1)^2 \tag{3.89}$$

The method of trees

In this section we shall discuss the method of trees, a method for constructing orthonormal sets of hyperspherical harmonics. The method was developed by a number of Russian authors [176, 177, 184, 202, 230, 277, 282, 296, 297] and by Prof. V. Aquilanti and his co-workers at the University of Perugia [12, 13, 70] .

We said above that an orthonormal set of 4-dimensional hyperspherical harmonics can be constructed according to the chain of subgroups $SO(4) \supset SO(3) \supset SO(2)$ or according to the alternative chain, $SO(4) \supset SO(2) \times SO(2)$. These two alternative ways of constructing an orthonormal set can be symbolized by the two *trees* shown in Figure 3.1.

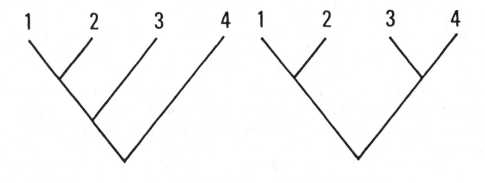

Figure 3.1

The meaning of these two alternative trees is as follows: If we wish to construct a set of hyperspherical harmonics according to the first tree (the *standard tree*), we begin with the harmonic polynomials

$$h_m(x_1, x_2) = (x_1 \pm ix_2)^m \qquad (3.90)$$

in the 2-dimensional space whose Cartesian coordinates are x_1 and x_2. Then

$$f_{l,m}(x_1, x_2, x_3) = x_3^{l-m} h_m(x_1, x_2) \tag{3.91}$$

will be a set of homogeneous polynomials of order l in the 3-dimensional space whose coordinates are x_1, x_2, and x_3. Equation (3.23) tells us that if we act on $f_l(x_1, x_2, x_3)$ with the operator

$$\begin{aligned}
O_l &= 1 - \frac{(x_1^2 + x_2^2 + x_3^2)}{2(2l-1)} \sum_{j=1}^{3} \frac{\partial^2}{\partial x_j^2} \\
&+ \frac{(x_1^2 + x_2^2 + x_3^2)^2}{8(2l-1)(2l-3)} \left(\sum_{j=1}^{3} \frac{\partial^2}{\partial x_j^2} \right)^2 - \dots
\end{aligned} \tag{3.92}$$

we will project out a set of harmonic polynomials $h_{l,m}(x_1, x_2, x_3)$. (If we divide these harmonic polynomials by $(x_1^2 + x_2^2 + x_3^2)^{l/2}$ and normalize them, we will have obtained the familiar 3-dimensional spherical harmonics Y_{lm}.) Having found $h_{l,m}(x_1, x_2, x_3)$, we can construct the homogeneous polynomials

$$f_{\lambda,l,m}(x_1, ..., x_4) = x_4^{\lambda-l} h_{l,m}(x_1, x_2, x_3) \tag{3.93}$$

and we can once more use equation (3.23) to project out the harmonic polynomials of highest order contained in it, $h_{\lambda,l,m}(x_1, ..., x_4)$

The standard tree can be used to construct an orthonormal set of hyperspherical harmonics in a space of arbitrarily high dimension: Suppose that we have already constructed an harmonic polynomial $h_l(x_1, ..., x_{d-1})$ in a $(d-1)$-dimensional space. Then

$$f_{\lambda,l}(x_1, ..., x_d) = x_d^{\lambda-l} h_l(x_1, ..., x_{d-1}) \tag{3.94}$$

will be a homogeneous polynomial of order λ in the d-dimensional space obtained by adding the final coordinate. Equation (3.23) tells us that if we act on this homogeneous polynomial with the operator

$$O_\lambda = \sum_{k=0}^{[\lambda/2]} \frac{(-1)^k (d + 2\lambda - 2k - 4)!!}{(2k)!!(d + 2\lambda - 4)!!} r^{2k} \Delta^k \tag{3.95}$$

we will project out from it an harmonic polynomial which will also be of order λ. The action of Δ^k on $f_{\lambda,l}$ is easy to calculate because

$$\Delta h_l(x_1, ..., x_{d-1}) = 0 \qquad (3.96)$$

from which it follows that

$$\Delta^k[x_d^{\lambda-l}h_l(x_1, ..., x_{d-1})] = \frac{(\lambda - l)!}{(\lambda - l - 2k)!}x_d^{\lambda-l-2k}h_l(x_1, ..., x_{d-1}) \qquad (3.97)$$

Combining (3.94) (3.95) and (3.97), we have

$$O_\lambda \left[x_d^{\lambda-l}h_l(x_1, ..., x_{d-1}) \right]$$
$$= r^{\lambda-l}h_l(x_1, ..., x_{d-1})\frac{(\lambda - l)!}{(d + 2\lambda - 4)!!}$$
$$\times \sum_{k=0}^{[(\lambda-l)/2]} \frac{(-1)^k(d + 2\lambda - 2k - 4)!!}{(2k)!!(\lambda - l - 2k)!}\left(\frac{x_d}{r}\right)^{\lambda-l-2k} \qquad (3.98)$$

Looking carefully at the series in equation (3.98), we can see that it can be expressed in terms of a Gegenbauer polynomial, since (from equation (3.69))

$$C_n^{(d'-2)/2}\left(\frac{x_d}{r}\right) = \frac{1}{(d' - 4)!!}\sum_{k=0}^{[n/2]}\frac{(-1)^k(d' + 2n - 2k - 4)!!}{(2k)!!(n - 2k)!}\left(\frac{x_d}{r}\right)^{n-2k} \qquad (3.99)$$

Comparing this with the series in equation (3.98), we can see that we ought to let $n = \lambda - l$ and $d' = d + 2l$. With these substitutions, (3.99) becomes:

$$C_{\lambda-l}^{\alpha+l}\left(\frac{x_d}{r}\right) = \frac{1}{(d + 2l - 4)!!}\sum_{k=0}^{[(\lambda-l)/2]}\frac{(-1)^k(d + 2\lambda - 2k - 4)!!}{(2k)!!(\lambda - l - 2k)!}\left(\frac{x_d}{r}\right)^{\lambda-l-2k} \qquad (3.100)$$

Combining equations (3.98) and (3.100), we obtain the relationship

$$h_{\lambda,l}(\mathbf{x}) = \mathcal{N}_{\lambda,l}r^{\lambda-l}C_{\lambda-l}^{\alpha+l}\left(\frac{x_d}{r}\right)h_l(x_1, ..., x_{d-1}) \qquad (3.101)$$

where $\alpha \equiv (d - 2)/2$, and where $\mathcal{N}_{\lambda,l}$ is a constant. If $h_l(x_1, ..., x_{d-1})$ is already normalized, then the harmonic polynomial $h_{\lambda,l}(\mathbf{x})$ will also be properly normalized if the constant $\mathcal{N}_{\lambda,l}$ is chosen to be

$$\mathcal{N}_{\lambda,l} = \left[\frac{\Gamma\left(\frac{d+2l}{2}\right)(d+2l-3)!(2\lambda+d-2)(\lambda-l)!}{\sqrt{\pi}\,\Gamma\left(\frac{d+2l-1}{2}\right)(d+2l-2)(\lambda+l+d-3)!} \right]^{1/2} \quad (3.102)$$

rather than the constant which results from projection. Equation (3.101) allows us to generate an orthonormal set of hyperspherical harmonics in a space of any dimension; and indeed this is the method which was used to construct the standard 4-dimensional hyperspherical harmonics shown in equation (2.16). However, in physical applications, the standard hyperspherical harmonics are not necessarily the most convenient ones to use. Therefore we shall now turn our attention to non-standard trees, an example of which is the second tree shown in Figure 2.1. To make the argument as general as possible, let us consider a general fork joining two subspaces whose dimensions are respectively d_1 and d_2, where $d_1 + d_2 = d$. Suppose that we have constructed the harmonic polynomials $h_{l_1}(\mathbf{x}_1)$ and $h_{l_2}(\mathbf{x}_2)$ in the two subspaces. We would like to use them as building blocks for constructing an harmonic polynomial of order λ in the d-dimensional space whose position vector is

$$\mathbf{x} = (\mathbf{x}_1, \mathbf{x}_2) \quad (3.103)$$

If we let

$$\begin{aligned}
r_1^2 &= \mathbf{x}_1 \cdot \mathbf{x}_1 \\
r_2^2 &= \mathbf{x}_2 \cdot \mathbf{x}_2 \\
r^2 &= \mathbf{x} \cdot \mathbf{x}
\end{aligned} \quad (3.104)$$

then

$$f_{\lambda,l_1,l_2}(\mathbf{x}) = r_1^{\beta} r_2^{\beta'} h_{l_1}(\mathbf{x}_1) h_{l_2}(\mathbf{x}_2) \quad (3.105)$$

will be a homogeneous polynomial of order λ in the coordinates $x_1, ..., x_d$, provided that β and β' are even and $\beta + \beta + l_1 + l_2 = \lambda$. If we act on $f_{\lambda,l_1,l_2}(\mathbf{x})$ with the projection operator shown in equation (3.95), we will obtain an harmonic polynomial of order λ. The simplest case is that for which $\beta = \beta' = 0$ and $l_1 + l_2 = \lambda$. In this simple case, the homogeneous polynomial shown in equation (3.105) is already harmonic because

$$\Delta[h_{l_1}(\mathbf{x}_1) h_{l_2}(\mathbf{x}_2)] = h_{l_2}(\mathbf{x}_2) \Delta_1 h_{l_1}(\mathbf{x}_1) + h_{l_1}(\mathbf{x}_1) \Delta_2 h_{l_2}(\mathbf{x}_2) = 0 \quad (3.106)$$

where

$$\Delta = \Delta_1 + \Delta_2$$

$$\Delta_1 = \sum_{j=1}^{d_1} \frac{\partial^2}{\partial x_j^2}$$

$$\Delta_2 = \sum_{j=d_1+1}^{d} \frac{\partial^2}{\partial x_j^2} \tag{3.107}$$

The next simplest case is that for which $\beta = 2$, $\beta' = 0$, and $l_1 + l_2 + 2 = \lambda$, so that

$$f_{\lambda,l_1,l_2}(\mathbf{x}) = r_1^2 h_{l_1}(\mathbf{x}_1) h_{l_2}(\mathbf{x}_2) \tag{3.108}$$

Making use of equation (3.12), we have:

$$\Delta[r_1^2 h_{l_1}(\mathbf{x}_1) h_{l_2}(\mathbf{x}_2)] = 2(d_1 + 2l_1) h_{l_1}(\mathbf{x}_1) h_{l_2}(\mathbf{x}_2) \tag{3.109}$$

and

$$\Delta^2[r_1^2 h_{l_1}(\mathbf{x}_1) h_{l_2}(\mathbf{x}_2)] = 0 \tag{3.110}$$

so that

$$h_{\lambda,l_1,l_2}(\mathbf{x}) = \left[r_1^2 - \frac{(d_1 + 2l_1)}{(d + 2\lambda - 4)} r^2 \right] h_{l_1}(\mathbf{x}_1) h_{l_2}(\mathbf{x}_2) \tag{3.111}$$

This can be rewritten in a more symmetrical form:

$$h_{\lambda,l_1,l_2}(\mathbf{x}) = \left[\frac{(d_2 + 2l_2)r_1^2 - (d_1 + 2l_1)r_2^2}{(d + 2\lambda - 4)} \right] h_{l_1}(\mathbf{x}_1) h_{l_2}(\mathbf{x}_2) \tag{3.112}$$

from which we can see that no new linearly independent harmonic polynomial would have been generated if we had let $\beta = 0$ and $\beta' = 2$.

The reader might now be wondering why we do not also try a homogeneous polynomial of the form

$$f_{\lambda,l_1,l_2}(\mathbf{x}) = x_j^{\lambda - l_1 - l_2} h_{l_1}(\mathbf{x}_1) h_{l_2}(\mathbf{x}_2) \tag{3.113}$$

since we could certainly project out from this an harmonic polynomial of order λ. Our reason for not doing so is that we are considering the general fork $SO(d) \supset SO(d_1) \times SO(d_2)$. The indices l_1 and l_2 label irreducible representations of $SO(d_1)$ and $SO(d_2)$ respectively, and this property would in general be lost if we multiplied by $x_j^{\lambda-l_1-l_2}$, but it is retained when we multiply by $r_1^{\beta} r_2^{\beta'}$. The powers of β and β' must be even in order that we should stay within the domain of polynomials in $x_1, ..., x_d$, unless $d_1 = 1$ or $d_2 = 1$.

To illustrate our discussion of the general fork, let us consider the alternative 4-dimensional hyperspherical harmonics constructed by means of the second tree shown in Figure 2.1. When $\lambda = 0$, the only possible harmonic polynomial is just a constant, and this fits with the predicted degeneracy, $(\lambda + 1)^2 = 1^2 = 1$. When $\lambda = 1$, there are two linearly independent harmonic polynomials with $l_1 = 1$ and $l_2 = 0$, namely $x_1 \pm ix_2$, and two with $l_1 = 0$ and $l_2 = 1$, namely $x_3 \pm ix_4$. This also checks with the predicted degeneracy $(\lambda + 1)^2 = 2^2 = 4$. When $\lambda = 2$, there are two harmonic polynomials with $l_1 = 2$ and $l_2 = 0$: $(x_1 \pm ix_2)^2$, two with $l_1 = 0$ and $l_2 = 2$: $(x_3 \pm ix_4)^2$, four with $l_1 = 1$ and $l_2 = 1$: $(x_1 \pm ix_2)(x_3 \pm ix_4)$, and one with $l_1 = 0$, $l_2 = 0$ and $\beta = 2$: $x_1^2 + x_2^2 - x_3^2 - x_4^2$. This makes a total of 9 linearly independent harmonic polynomials of order $\lambda = 2$, which checks with the predicted degeneracy, $(\lambda + 1)^2 = 3^2 = 9$. We can find the corresponding hyperspherical harmonics by dividing by r^2 and normalizing, as illustrated in Table 4.1. As we shall see, the alternative hyperspherical harmonics shown in this table correspond in Fock's mapping to the Fourier transforms of the orbitals which result when the hydrogenlike Schrödinger equation is separated in paraboloidal coordinates, and the quantum numbers labeling the hyperspherical harmonics in Table 4.1 are those which are often used to label the direct-space paraboloidal orbitals.

One can show that for a general fork between subspaces of dimension d_1 and d_2, an harmonic polynomial of order λ can be written in the form:

$$h_{\lambda,l_1,l_2}(\mathbf{x}) \sim r^{\lambda-l_1-l_2} P_{(\lambda-l_1-l_2)/2}^{(l_2+\alpha_2,l_1+\alpha_1)}\left(\frac{r_1^2 - r_2^2}{r^2}\right) h_{l_1}(\mathbf{x}_1) h_{l_2}(\mathbf{x}_2) \quad (3.114)$$

where $\alpha_j \equiv (d_j - 2)/2$ and where $P_n^{(a,b)}(x)$ is a Jacobi polynomial. When

$d_1 = d_2 = 2$, $\alpha_1 = \alpha_2 = 0$, and we have

$$h_{\lambda,l_1,l_2}(\mathbf{x}) \sim r^{\lambda-l_1-l_2} P^{(l_2,l_1)}_{(\lambda-l_1-l_2)/2}\left(\frac{r_1^2 - r_2^2}{r^2}\right) h_{l_1}(\mathbf{x}_1) h_{l_2}(\mathbf{x}_2) \qquad (3.115)$$

For example, when $\lambda = 3$, $l_1 = 1$ and $l_2 = 0$, equation (3.115) becomes:

$$\begin{aligned}
h_{3,1,0}(\mathbf{x}) &\sim r^2 P_1^{(0,1)}\left(\frac{r_1^2 - r_2^2}{r^2}\right) h_1(\mathbf{x}_1) \\
&\sim (2r_2^2 - 4r_1^2) h_1(\mathbf{x}_1) \qquad (3.116)
\end{aligned}$$

which checks with equation (3.112). For more details concerning these relationships, the interested reader is referred to the Russian literature and to the papers of Aquilanti and his coworkers.

Table 3.1
Decomposition of monomials into harmonic polynomials
The indices i, j, k are assumed to be all unequal.
The terms in brackets are harmonic.

n	$m_n = h_n + r^2 h_{n-2} + \ldots$
1	$x_i = (x_i)$
2	$x_i^2 = \left(x_i^2 - \dfrac{r^2}{d} \right) + r^2 \left(\dfrac{1}{d} \right)$ $x_i x_j = (x_i x_j)$
3	$x_i^3 = \left(x_i^3 - \dfrac{3r^2 x_i}{d+2} \right) + r^2 \left(\dfrac{3x_i}{d+2} \right)$ $x_i^2 x_j = \left(x_i^2 x_j - \dfrac{r^2 x_j}{d+2} \right) + r^2 \left(\dfrac{x_j}{d+2} \right)$ $x_i x_j x_k = (x_i x_j x_k)$

Table 3.2

Gegenbauer polynomials

$$(\alpha)_j \equiv \alpha(\alpha+1)(\alpha+2)...(\alpha+j-1)$$

λ	$C_\lambda^\alpha(\zeta)$	$C_\lambda^1(\zeta)$
0	1	1
1	$2\alpha\zeta$	2ζ
2	$2(\alpha)_2\zeta^2 - \alpha$	$4\zeta^2 - 1$
3	$\frac{1}{3}\left[4(\alpha)_3\zeta^3 - 6(\alpha)_2\zeta\right]$	$8\zeta^3 - 4\zeta$
4	$\frac{1}{6}\left[4(\alpha)_4\zeta^4 - 12(\alpha)_3\zeta^2 + 3(\alpha)_2\right]$	$16\zeta^4 - 12\zeta^2 + 1$
5	$\frac{1}{15}\left[4(\alpha)_5\zeta^5 - 20(\alpha)_4\zeta^3 + 15(\alpha)_3\zeta\right]$	$32\zeta^5 - 32\zeta^3 + 6\zeta$

Table 3.3

This table shows the dependence of the total generalized solid angle, $I(0)$, and the degeneracy of the hyperspherical harmonics, ω, on the dimension, d. α and K_λ are constants which appear in the sum rule for hyperspherical harmonics, equation (3.80).

d	α	$I(0)$	K_λ	ω
2	0	2π	0	2
3	$\dfrac{1}{2}$	4π	$\dfrac{4\pi}{2\lambda+1}$	$2\lambda+1$
4	1	$2\pi^2$	$\dfrac{2\pi^2}{\lambda+1}$	$(\lambda+1)^2$
5	$\dfrac{3}{2}$	$\dfrac{8\pi^2}{3}$	$\dfrac{8\pi^2}{2\lambda+3}$	$\dfrac{(2\lambda+3)(\lambda+2)(\lambda+1)}{6}$

Exercises

1. Which of the following polynomials in a d-dimensional space are homogeneous? Which are harmonic? (r is the hyperradius.)

   (a) $x_1^3 + x_2^3$

   (b) $x_1^3 + x_2^2$

   (c) $2x_1^3 + x_2^3$

   (d) $x_1^3 - x_2^3$

   (e) $x_1^3 - 3x_1 r^2/(d+2)$

   (f) $x_1^2 x_2 x_3 - r^2 x_2 x_3/(d+4)$

2. Use equation (3.24) to find expressions analogous to (3.18) for the harmonic decomposition of a 4th-order polynomial, f_4.

3. Use equation (3.49) to calculate the normalization factor, in a 4-dimensional space, for the hyperspherical harmonic shown in equation (3.6). Compare this result with Table 2.4.

4. Show that the hyperspherical harmonics with $\lambda = 1$ in Table 2.3 fulfill the sum rule of equation (3.72). Show that they are properly normalized.

5. Use equation (3.72) to show that

$$\frac{2\lambda + d - 2}{(d-2)I(0)} \int d\Omega \ C_\lambda^\alpha(\hat{\mathbf{x}} \cdot \hat{\mathbf{x}}')C_{\lambda'}^\alpha(\hat{\mathbf{x}} \cdot \hat{\mathbf{x}}'') = \delta_{\lambda'\lambda}C_\lambda^\alpha(\hat{\mathbf{x}}' \cdot \hat{\mathbf{x}}'')$$

Chapter 4

THE MOMENTUM-SPACE WAVE EQUATION

The d-dimensional Schrödinger equation

In direct space, the Schrödinger equation is a differential equation; but in momentum-space it becomes an integral equation. We will begin by deriving this integral equation for a system with d Cartesian coordinates. For an N-particle system, $d = 3N$; while for a single particle moving in an external field, $d = 3$. If we let

$$e^{i\mathbf{p}\cdot\mathbf{x}} \equiv e^{i(p_1 x_1 + \dots + p_d x_d)} \tag{4.1}$$

represent a d-dimensional plane wave, then the wave function and its Fourier transform are related by:

$$\psi(\mathbf{x}) = \frac{1}{(2\pi)^{d/2}} \int dp \; e^{i\mathbf{p}\cdot\mathbf{x}} \psi^t(\mathbf{p})$$

$$\psi^t(\mathbf{p}) = \frac{1}{(2\pi)^{d/2}} \int dx \; e^{-i\mathbf{p}\cdot\mathbf{x}} \psi(\mathbf{x}) \tag{4.2}$$

The wave function and its Fourier transform are functions of d coordinates in direct and reciprocal space respectively; and the volume elements in direct and reciprocal space also involve d coordinates. We represent this using the following notation:

$$\psi(\mathbf{x}) \equiv \psi(x_1, \dots, x_d)$$

59

$$\psi^t(\mathbf{p}) \equiv \psi^t(p_1, \dots, p_d)$$
$$dx \equiv dx_1 dx_2 \dots dx_d$$
$$dp \equiv dp_1 dp_2 \dots dp_d \tag{4.3}$$

As we mentioned in Chapter 1, the direct-space Schrödinger equation can then be written in the form:

$$\left[-\Delta + p_0^2 + 2V(\mathbf{x})\right]\psi(\mathbf{x}) = 0 \tag{4.4}$$

where we have used mass-weighted coordinates and atomic units, with the notation:

$$p_0^2 \equiv -2E \tag{4.5}$$

and

$$\Delta \equiv \sum_{j=1}^{d} \frac{\partial^2}{\partial x_j^2} \tag{4.6}$$

If we substitute the expression for $\psi(\mathbf{x})$ in terms of its Fourier transform into (4.4), we have:

$$\int dp\, e^{-i\mathbf{p}\cdot\mathbf{x}} \left[p^2 + p_0^2 + 2V(\mathbf{x})\right]\psi^t(\mathbf{p}) = 0 \tag{4.7}$$

since $-\Delta$, acting on the plane wave, brings down a factor of p^2. If we multiply (4.7) on the left by $e^{-i\mathbf{p}'\cdot\mathbf{x}}$, and integrate over dx, we obtain:

$$\int dp\, (p^2 + p_0^2)\psi^t(\mathbf{p}) \int dx e^{i(\mathbf{p}-\mathbf{p}')\cdot\mathbf{x}}$$
$$+2\int dp \int dx\, e^{i(\mathbf{p}-\mathbf{p}')\cdot\mathbf{x}}V(\mathbf{x})\psi^t(\mathbf{p}) \quad = \quad 0 \tag{4.8}$$

However,

$$\int dx e^{i(\mathbf{p}-\mathbf{p}')\cdot\mathbf{x}} = (2\pi)^d \delta(\mathbf{p}-\mathbf{p}') \tag{4.9}$$

so that (4.8) becomes:

$$(2\pi)^d \int dp\, \delta(\mathbf{p}-\mathbf{p}')(p^2 + p_0^2)\psi^t(\mathbf{p})$$
$$+2\int dp \int dx\, e^{i(\mathbf{p}-\mathbf{p}')\cdot\mathbf{x}}V(\mathbf{x})\psi^t(\mathbf{p}) \quad = \quad 0 \tag{4.10}$$

Using the Dirac δ-function to integrate the first term over dp, we obtain:

$$(p'^2 + p_0^2)\psi^t(\mathbf{p'}) = -\frac{2}{(2\pi)^{d/2}} \int dp \, V^t(\mathbf{p'} - \mathbf{p})\psi^t(\mathbf{p}) \qquad (4.11)$$

where

$$V^t(\mathbf{p'} - \mathbf{p}) \equiv \frac{1}{(2\pi)^{d/2}} \int dx \, e^{-i(\mathbf{p'} - \mathbf{p}) \cdot \mathbf{x}} V(\mathbf{x}) \qquad (4.12)$$

Equation (4.11) is the momentum-space Schrödinger equation.

Hydrogenlike orbitals in p-space

Having discussed the properties of Gegenbauer polynomials, and having derived equation (4.11), we are now in a position to look in more detail at Fock's momentum-space treatment of hydrogenlike atoms, the results of which were shown in equations (2.13)-(2.18). When $d = 3$, $p_0 \to k_\mu$, $V(\mathbf{x}) \to v(\mathbf{x})$ and

$$v(\mathbf{x}) = -\frac{Z}{r} \qquad (4.13)$$

then

$$v^t(\mathbf{p'} - \mathbf{p}) = -\frac{Z}{(2\pi)^{3/2}} \int d^3x \, \frac{j_0(|\mathbf{p'} - \mathbf{p}|r)}{r} = -\sqrt{\frac{2}{\pi}} \frac{Z}{|\mathbf{p'} - \mathbf{p}|^2} \qquad (4.14)$$

If we substitute (4.14) into (4.11), letting $d = 3$, we obtain the p-space Schrödinger equation for hydrogenlike atoms:

$$(p'^2 + k_\mu^2)\psi^t(\mathbf{p'}) = \frac{Z}{\pi^2} \int d^3p \, \frac{1}{|\mathbf{p'} - \mathbf{p}|^2}\psi^t(\mathbf{p}) \qquad (4.15)$$

Following Fock's method, we now project p-space onto the surface of a 4-dimensional hypersphere by means of the transformation:

$$u_j = \frac{2k_\mu p_j}{k_\mu^2 + p^2} \qquad j = 1, 2, 3$$

$$u_4 = \frac{k_\mu^2 - p^2}{k_\mu^2 + p^2} \qquad (4.16)$$

The components of the vector $\mathbf{p}$ are related to its magnitude, p, through the angles θ_p, and ϕ_p:

$$\begin{aligned}
p_1 &= p\sin\theta_p\cos\phi_p \\
p_2 &= p\sin\theta_p\sin\phi_p \\
p_3 &= p\cos\theta_p
\end{aligned}$$

(4.17)

and we can define an additional angle, χ, by

$$\begin{aligned}
\frac{2k_\mu p}{k_\mu^2 + p^2} &= \sin\chi \\
\frac{k_\mu^2 - p^2}{k_\mu^2 + p^2} &= \cos\chi
\end{aligned}$$

(4.18)

Then, expressed in terms of these angles, the unit vectors in $u_1, ..., u_4$ become:

$$\begin{aligned}
u_1 &= \sin\chi\sin\theta_p\cos\phi_p \\
u_2 &= \sin\chi\sin\theta_p\sin\phi_p \\
u_3 &= \sin\chi\cos\theta_p \\
u_4 &= \cos\chi
\end{aligned}$$

(4.19)

From equations (4.17)-(4.19) it follows that the volume element in p-space,

$$d^3p = p^2 dp\ \sin\theta_p\ d\theta_p d\phi_p$$

(4.20)

is related to the generalized solid angle element on the hypersphere by

$$d\Omega = \sin^2\chi\sin\theta_p\ d\chi d\theta_p d\phi_p = \left(\frac{2k_\mu}{k_\mu^2 + p^2}\right)^3 d^3p$$

(4.21)

since, from (4.18),

$$\frac{d\chi}{dp} = \frac{2k_\mu}{k_\mu^2 + p^2}$$

(4.22)

With a little work, one can also derive the relationship [30],

$$\frac{1}{|\mathbf{p}' - \mathbf{p}|^2} = \frac{4k_\mu^2}{(k_\mu^2 + p^2)(k_\mu^2 + p'^2)}\frac{1}{|\mathbf{u}' - \mathbf{u}|^2}$$

(4.23)

If we substitute (4.23) and (4.21) into (4.15), the integral becomes

$$(p'^2 + k_\mu^2)^2 \psi^t(\mathbf{p}') = \frac{Z}{2k_\mu \pi^2} \int d\Omega \, \frac{(p^2 + k_\mu^2)^2}{|\mathbf{u}' - \mathbf{u}|^2} \psi^t(\mathbf{p}) \tag{4.24}$$

If we now let

$$\psi^t(\mathbf{p}) = M(p)F(\mathbf{u}) \tag{4.25}$$

where

$$M(p) = \frac{4k_\mu^{5/2}}{(k_\mu^2 + p^2)^2} \tag{4.26}$$

the integral equation, (4.24), takes on a simpler form:

$$F(\mathbf{u}') = \frac{Z}{2k_\mu \pi^2} \int d\Omega \, \frac{1}{|\mathbf{u}' - \mathbf{u}|^2} F(\mathbf{u}) \tag{4.27}$$

We can now use equation (3.65) to expand the kernel of the integral equation in terms of Gegenbauer polynomials: If we let

$$r_> = r_< = 1 \tag{4.28}$$

and $\alpha = 1$ (since we have a 4-dimensional hypersphere), and if we make use of the sum rule, (3.72), then (3.65) becomes:

$$\frac{1}{|\mathbf{u}' - \mathbf{u}|^2} = \sum_{\lambda=0}^{\infty} C_\lambda^1(\mathbf{u} \cdot \mathbf{u}') = \sum_{\lambda=0}^{\infty} \frac{2\pi^2}{\lambda + 1} \sum_{\{\mu\}} Y_{\lambda\{\mu\}}(\mathbf{u}') Y_{\lambda\{\mu\}}^*(\mathbf{u}) \tag{4.29}$$

If we substitute this expansion into (4.27), we obtain:

$$F(\mathbf{u}') = \sum_{\lambda=0}^{\infty} \frac{Z}{k_\mu(\lambda + 1)} \sum_{\{\mu\}} Y_{\lambda\{\mu\}}(\mathbf{u}') \int d\Omega \, Y_{\lambda\{\mu\}}^*(\mathbf{u}) F(\mathbf{u}) \tag{4.30}$$

Because of the orthonormality of the hyperspherical harmonics,

$$\int d\Omega \, Y_{\lambda\{\mu\}}^*(\mathbf{u}) Y_{\lambda'\{\mu'\}}(\mathbf{u}) = \delta_{\lambda'\lambda} \delta_{\{\mu'\},\{\mu\}} \tag{4.31}$$

(4.30) will be fulfilled provided that

$$F(\mathbf{u}) = Y_{\lambda'\{\mu'\}}(\mathbf{u}) \tag{4.32}$$

and provided that

$$\frac{Z}{k_\mu(\lambda' + 1)} = 1 \tag{4.33}$$

If we make the identification: $\lambda' + 1 \equiv n$, then equations (4.25) and (4.32) give us Fock's momentum-space hydrogenlike orbitals:

$$\psi^t(\mathbf{p}) = M(p)F(\mathbf{u}) = M(p)Y_{n-1,\{\mu\}}(\mathbf{u}) \tag{4.34}$$

while (4.33) tells us the corresponding energies:

$$\epsilon = -\frac{k_\mu^2}{2} = -\frac{Z^2}{2n^2} \qquad n = 1, 2, 3, ... \tag{4.35}$$

It is interesting to notice that we do not need to make the identification: $\{\mu\} \equiv \{l, m\}$, which would correspond to 4-dimensional hyperspherical harmonics constructed by means of the subgroup chain,

$$SO(4) \supset SO(3) \supset SO(2) \tag{4.36}$$

Any alternative way of constructing an orthonormal set of 4-dimensional hyperspherical harmonics would yield an equally good set of momentum-space hydrogenlike orbitals. For example, as Aquilanti and his coworkers have shown [9-27, 70], the 4-dimensional hyperspherical harmonics in Table 4.1 correspond to the set of momentum-space hydrogenlike orbitals whose direct-space counterparts can be found by separating the Schrödinger equation in parabolic coordinates:

$$\begin{aligned} \xi &\equiv r + z \\ \eta &\equiv r - z \\ \phi &\equiv \tan^{-1}\left(\frac{y}{x}\right) \end{aligned} \tag{4.37}$$

The the direct-space orbitals have the form [204]:

$$w_{n_1,n_2,m} = \sqrt{\frac{n_1!n_2!k_\mu^{2|m|+3}}{\pi n[(n_1 + |m|)!(n_2 + |m|)!]^3}}$$
$$\times e^{-k_\mu(\xi+\eta)/2}(\xi\eta)^{|m|/2}e^{im\phi}L_{n_1+|m|}^{|m|}(k_\mu\xi)L_{n_2+|m|}^{|m|}(k_\mu\eta) \tag{4.38}$$

where $n = n_1 + n_2 + m + 1$ and m play their usual roles as the principal quantum number and the magnetic quantum number respectively, and where the functions $L_p^q(\rho)$ are the associated Laguerre polynomials (Table 4.2):

$$L_p^q(\rho) = \frac{d^q}{d\rho^q}\left[e^\rho \frac{d^p}{d\rho^p}\left(e^{-\rho}\rho^p\right)\right] \qquad (4.39)$$

The first few direct-space orbitals in parabolic coordinates are shown in Table 4.3. It is interesting to see that the hyperspherical harmonics to which they correspond (Table 4.1) are constructed according to the subgroup chain

$$SO(4) \supset SO(2) \times SO(2) \qquad (4.40)$$

Momentum-space orthonormality of Sturmian basis functions

Since we intend to use momentum-space hydrogenlike orbitals as basis functions for solving general problems in quantum chemistry, it is essential to investigate their orthonormality relations. Let us first look at the case where 3-dimensional p-space is projected onto the surface of a 4-dimensional hypersphere, as shown in equations (4.13)-(4.35). The momentum-space hydrogenlike orbitals are related to the harmonics on this 4-dimensional hypersphere by

$$\chi_{n,l,m}^t(\mathbf{p}) = M(p)Y_{n-1,l,m}(\mathbf{u}) \equiv \frac{4k_\mu^{5/2}}{(k_\mu^2 + p^2)^2}Y_{n-1,l,m}(\mathbf{u}) \qquad (4.41)$$

Since the hyperspherical harmonics obey the orthonormality relation:

$$\int d\Omega \, Y_{n'-1,l',m'}^*(\mathbf{u})Y_{n-1,l,m}(\mathbf{u}) = \delta_{n',n}\delta_{l',l}\delta_{m',m} \qquad (4.42)$$

it follows that

$$\int d^3p \left(\frac{2k_\mu}{k_\mu^2 + p^2}\right)^3 M(p)^{-2}\chi_{n',l',m'}^{*t}(\mathbf{p})\chi_{n,l,m}^t(\mathbf{p}) = \delta_{n',n}\delta_{l',l}\delta_{m',m} \qquad (4.43)$$

or

$$\int d^3p \left(\frac{k_\mu^2 + p^2}{2k_\mu^2}\right) \chi_{n',l',m'}^{*t}(\mathbf{p})\chi_{n,l,m}^t(\mathbf{p}) = \delta_{n',n}\delta_{l',l}\delta_{m',m} \qquad (4.44)$$

Thus, provided that we keep k_μ constant for all its members, the set of momentum-space hydrogenlike orbitals obey a weighted orthonormality relation, the weighting factor being $(k_\mu^2 + p^2)/(2k_\mu^2)$. As we mentioned in the introduction, one-particle hydrogenlike Sturmian basis sets of this type were introduced by Shull and Löwdin, and they are widely used in atomic physics and quantum chemistry. The members of such a set can be thought of as solutions to the Schrödinger equation of a hydrogen atom with the nuclear charge weighted in such a way that all of the members of the set correspond to the same value of the energy, regardless of their quantum numbers. The constancy of the energy within the basis set is essential for the validity of equations (4.43) and (4.44).

Not only do the members of a momentum-space Sturmian basis set obey a weighted orthonormality relation in p-space - but also their direct-space counterparts obey a weighted orthonormality relation; and we will now show that the weighting factor in direct space is proportional to the potential. In order to show this, we recall two theorems of Fourier analysis. The first theorem states that if if $f(\mathbf{x})$ and $g(\mathbf{x})$ are any two functions in direct space, and if $f^t(\mathbf{p})$ and $g^t(\mathbf{p})$ are their Fourier transforms, then

$$\int dx \; f^*(\mathbf{x})g(\mathbf{x}) = \int dp \; f^{t*}(\mathbf{p})g^t(\mathbf{p}) \qquad (4.45)$$

This theorem follows from (4.9), since

$$\int dx \; f^*(\mathbf{x})g(\mathbf{x})$$
$$= \frac{1}{(2\pi)^d} \int dp \int dp' \; f^{t*}(\mathbf{p})g^t(\mathbf{p}') \int dx \; e^{i(\mathbf{p}'-\mathbf{p})\cdot\mathbf{x}}$$
$$= \int dp \int dp' \; f^{t*}(\mathbf{p})g^t(\mathbf{p}')\delta(\mathbf{p}' - \mathbf{p})$$
$$= \int dp \; f^{t*}(\mathbf{p})g^t(\mathbf{p}) \qquad (4.46)$$

We shall also need the Fourier convolution theorem [27]:

$$\int dx \; e^{-i\mathbf{p}\cdot\mathbf{x}} f(\mathbf{x})g(\mathbf{x}) = \int dp' \; f^t(\mathbf{p}')g^t(\mathbf{p} - \mathbf{p}') \qquad (4.47)$$

which can be proved in a similar way. From the Fourier convolution theorem, (4.47), it follows that the momentum-space Schrödinger equation, (4.11), can be rewritten in the form:

$$(p_0^2 + p^2)\psi^t(\mathbf{p}) = -\frac{2}{(2\pi)^{d/2}} \int dx \ e^{-i\mathbf{p}\cdot\mathbf{x}} V(\mathbf{x})\psi(\mathbf{x}) \qquad (4.48)$$

If we let $d \to 3$, $p_0 \to k_\mu$, $V(\mathbf{x}) \to -nk_\mu/r$ and and $\psi^t(\mathbf{p}) \to \chi_{n,l,m}^t(\mathbf{p})$, this becomes

$$(k_\mu^2 + p^2)\chi_{n,l,m}^t(\mathbf{p}) = \frac{2}{(2\pi)^{3/2}} \int d^3x \ e^{-i\mathbf{p}\cdot\mathbf{x}} \frac{nk_\mu}{r} \chi_{n,l,m}(\mathbf{x}) \qquad (4.49)$$

Equation (4.49) is the momentum-space counterpart of the direct-space equation (1.11):

$$\left[-\frac{1}{2}\Delta + \frac{1}{2}k_\mu^2 - \frac{nk_\mu}{r} \right] \chi_{n,l,m}(\mathbf{x}) = 0 \qquad (4.50)$$

Taking the complex conjugate of (4.49), we obtain:

$$\left[\frac{nk_\mu}{r} \chi_{n,l,m}^*(\mathbf{x}) \right]^{t*} = \frac{1}{2}(k_\mu^2 + p^2)\chi_{n,l,m}^{t*}(\mathbf{p}) \qquad (4.51)$$

The quantity in square brackets on the left-hand side of (4.51) can be identified with $f^t(\mathbf{p})$ in (4.46). If we let

$$g^t(\mathbf{p}) = \chi_{n',l',m'}^t(\mathbf{p}) \qquad (4.52)$$

then (4.44), (4.46) and (4.51) yield the direct-space orthonormality relation for the members of our hydrogenlike Sturmian basis set:

$$\frac{n}{k_\mu} \int d^3x \ \chi_{n,l,m}^*(\mathbf{x}) \frac{1}{r} \chi_{n',l',m'}(\mathbf{x}) = \delta_{n,n'}\delta_{l',l}\delta_{m',m} \qquad (4.53)$$

Looking at equation (4.53), we can see that the direct-space orthonormality relation has a weighting factor which is proportional to the potential. We saw earlier, in equations (1.18)-(1.21), why this must be true in a general d-dimensional case, where the potential need not be a Coulomb potential. Members of any Sturmian basis set are orthonormal in direct space, with a weighting factor which is proportional to the potential. This still does not tell us how to normalize the basis set. However, if we normalize it according to the condition

$$\int dx \; \phi_{\nu'}^*(\mathbf{x})V_0(\mathbf{x})\phi_\nu(\mathbf{x}) = \frac{2E}{\beta_\nu}\delta_{\nu',\nu} = -\frac{p_0^2}{\beta_\nu}\delta_{\nu',\nu} \qquad (4.54)$$

then from (4.48), with $V(\mathbf{x}) \to \beta_\nu V_0(\mathbf{x})$ and $\psi \to \phi_\nu$, we have:

$$(p_0^2 + p^2)\phi_\nu^t(\mathbf{p}) = -2\beta_\nu \left[V_0\phi_\nu\right]^t (\mathbf{p}) \qquad (4.55)$$

Combining (4.46), (4.54) and (4.55) then yields:

$$\begin{aligned}
\int dx \; \phi_{\nu'}^*(\mathbf{x})V_0(\mathbf{x})\phi_\nu(\mathbf{x}) &= \int dp \; \phi_{\nu'}^{t*}(\mathbf{p})[V_0\phi_\nu]^t(\mathbf{p}) \\
&= -\int dp \; \phi_{\nu'}^{t*}(\mathbf{p}) \left(\frac{p_0^2 + p^2}{2\beta_\nu}\right) \phi_\nu^t(\mathbf{p}) \\
&= -\frac{p_0^2}{\beta_\nu}\delta_{\nu',\nu} \qquad (4.56)
\end{aligned}$$

so that the orthonormality relation in reciprocal space is:

$$\int dp \left(\frac{p_0^2 + p^2}{2p_0^2}\right) \phi_{\nu'}^{*t}(\mathbf{p})\phi_\nu^t(\mathbf{p}) = \delta_{\nu',\nu} \qquad (4.57)$$

The meaning of the orthonormality relations (4.54) and (4.57) requires a further comment: Since we are dealing with a many-dimensional space, ν represents a set of quantum numbers rather than a single quantum number. Orthogonality with respect to the quantum number (or numbers) on which β_ν depends follows automatically from (1.20), i.e. Sturmian basis functions corresponding to different values of β_ν are necessarily orthogonal; but orthogonality with respect to the minor quantum numbers must be constructed or proved in some other way, for example by symmetry arguments. The functional space spanned by a set of Sturmian basis functions is a Sobolev space; and the interested reader is referred to an excellent article by E. Weniger [301], which contains a discussion of the orthonormality and completeness relations of basis sets in Sobolev spaces.

Sturmian expansions of d-dimensional plane waves

As has been shown by Aquilanti et al. [27, 28], d-dimensional plane waves can be expanded in terms of Sturmian basis sets. If we expand the plane wave in terms of the complete set of momentum-space functions, $\phi_\nu^t(\mathbf{p})$:

$$e^{i\mathbf{p}\cdot\mathbf{x}} = \left(\frac{p_0^2 + p^2}{2p_0^2}\right)\sum_\nu \phi_\nu^{*t}(\mathbf{p})a_\nu \tag{4.58}$$

then we can use the weighted orthonormality relations, (4.57), to determine the expansion coefficients a_ν. Multiplying both sides of (4.58) by $\phi_{\nu'}^t(\mathbf{p})$ and integrating over reciprocal space, we obtain:

$$a_\nu = \int dp \, e^{i\mathbf{p}\cdot\mathbf{x}}\phi_\nu^t(\mathbf{p}) = (2\pi)^{d/2}\phi_\nu(\mathbf{x}) \tag{4.59}$$

so that

$$e^{i\mathbf{p}\cdot\mathbf{x}} = (2\pi)^{d/2}\left(\frac{p_0^2 + p^2}{2p_0^2}\right)\sum_\nu \phi_\nu^{*t}(\mathbf{p})\phi_\nu(\mathbf{x}) \tag{4.60}$$

Table 4.1

Hyperspherical harmonics corresponding to solutions for
the hydrogen atom in parabolic coordinates.

n_1	n_2	m	$\sqrt{2\pi}\,\mathcal{Y}_{n_1,n_2,m}(\mathbf{u})$
0	0	0	1
0	0	1	$-i\sqrt{2}(u_1 + iu_2)$
0	0	-1	$-i\sqrt{2}(u_1 - iu_2)$
1	0	0	$\sqrt{2}(u_4 + iu_3)$
0	1	0	$\sqrt{2}(u_4 - iu_3)$

Table 4.1 continued

n_1	n_2	m	$\sqrt{2}\pi\, \mathcal{Y}_{n_1,n_2,m}(\mathbf{u})$
0	0	2	$-\sqrt{3}(u_1 + iu_2)^2$
0	0	-2	$-\sqrt{3}(u_1 - iu_2)^2$
2	0	0	$\sqrt{3}(u_4 + iu_3)^2$
0	2	0	$\sqrt{3}(u_4 - iu_3)^2$
1	1	0	$\sqrt{3}(u_1^2 + u_2^2 - u_3^2 - u_4^2)$
1	0	1	$\sqrt{6}(u_4 + iu_3)(u_1 + iu_2)$
1	0	-1	$\sqrt{6}(u_4 + iu_3)(u_1 - iu_2)$
0	1	1	$\sqrt{6}(u_4 - iu_3)(u_1 + iu_2)$
0	1	-1	$\sqrt{6}(u_4 - iu_3)(u_1 - iu_2)$

Table 4.2

Associated Laguerre polynomials, L_p^q, (Equation (4.39)).

p	$q = 0$	$q = 1$	$q = 2$	$q=3$
0	1			
1	$1 - \rho$	-1		
2	$2 - 4\rho + \rho^2$	$-4 + 2\rho$	2	
3	$6 - 18\rho + 9\rho^2 - \rho^3$	$-18 + 18\rho - 3\rho^2$	$18 - 6\rho$	-6

Table 4.3

Hydrogenlike orbitals in parabolic coordinates
(Equation (4.38))
$$\mathbf{t} \equiv k_\mu \mathbf{x}, \qquad t \equiv k_\mu r$$

n_1	n_2	m	$\left(\pi/k_\mu^3\right)^{1/2} w_{n_1,n_2,m}(\mathbf{x})$
0	0	0	$e^{-k_\mu(\xi+\eta)/2} = e^{-t}$
0	0	1	$\frac{1}{\sqrt{2}} e^{-k_\mu(\xi+\eta)/2} k_\mu(\xi\eta)^{1/2} e^{i\phi} = \frac{1}{\sqrt{2}} e^{-t}(t_1 + it_2)$
0	0	-1	$\frac{1}{\sqrt{2}} e^{-k_\mu(\xi+\eta)/2} k_\mu(\xi\eta)^{1/2} e^{-i\phi} = \frac{1}{\sqrt{2}} e^{-t}(t_1 - it_2)$
1	0	0	$\frac{1}{\sqrt{2}} e^{-k_\mu(\xi+\eta)/2}(1 - k_\mu\xi) = \frac{1}{\sqrt{2}} e^{-t}(1 - t - t_3)$
0	1	0	$\frac{1}{\sqrt{2}} e^{-k_\mu(\xi+\eta)/2}(1 - k_\mu\eta) = \frac{1}{\sqrt{2}} e^{-t}(1 - t + t_3)$

Table 4.4
d-dimensional hydrogenlike Sturmians

n	l	$R_{n,l}(r)$ $t \equiv k_\mu r$
1	0	$\dfrac{(2k_\mu)^{d/2}}{\sqrt{(d-1)!}} e^{-t}$
2	0	$\dfrac{(2k_\mu)^{d/2}}{(d-2)!} \sqrt{\dfrac{(d-1)!}{d+1}} \left[1 - \dfrac{2t}{d-1}\right] e^{-t}$
2	1	$\dfrac{(2k_\mu)^{d/2}}{\sqrt{(d+1)!}} (2t) e^{-t}$
3	0	$\dfrac{(2k_\mu)^{d/2}}{(d-2)!} \sqrt{\dfrac{d!}{d+3}} \left[1 - \dfrac{4t}{d-1} + \dfrac{4t^2}{d(d-1)}\right] e^{-t}$
3	1	$\dfrac{(2k_\mu)^{d/2}}{d!} \sqrt{\dfrac{(d+1)!}{d+3}} (2t) \left[1 - \dfrac{2t}{d+1}\right] e^{-t}$
3	2	$\dfrac{(2k_\mu)^{d/2}}{\sqrt{(d+3)!}} (2t)^2 e^{-t}$

Exercises

1. Use equations (4.17)-(4.20) to derive equation (4.21).

2. Use equation (4.39) to generate the associated Laguerre polynomials of Table 4.2.

3. From equation (4.38) and the associated Laguerre polynomials in Table 4.2, calculate the parabolic hydrogenlike orbitals shown in Table 4.3. Express these functions as linear combinations of $\chi_{nlm}(\mathbf{x})$.

Chapter 5

MANY-CENTER POTENTIALS

Expansion of the kernel

In Chapter 1 we very briefly discussed the many-center problem, and we mentioned that it is convenient to solve the many-center one-electron equation, (1.55), in momentum space, making use of the theory of hyperspherical harmonics. We are now in a position to discuss in detail how this problem may be solved. Shibuya and Wulfman [272] were able to extend Fock's method of treating hydrogenlike atoms to many-center Coulomb potentials; and their treatment has been further developed by a number of other authors. The method relies on a generalization of Fock's expansion of the kernel of the momentum-space Schrödinger equation for hydrogenlike atoms. This generalization is made in the following way: Suppose that $\mathbf{x} = \{x, y, z\}$ and that

$$v(\mathbf{x}) = -\sum_a \frac{Z_a}{|\mathbf{x} - \mathbf{X}_a|} \tag{5.1}$$

Then from equation (4.14) we have

$$v^t(\mathbf{p} - \mathbf{p}') = -\sqrt{\frac{2}{\pi}} \frac{1}{|\mathbf{p}' - \mathbf{p}|^2} \sum_a Z_a e^{-i(\mathbf{p} - \mathbf{p}') \cdot \mathbf{X}_a} \tag{5.2}$$

By combining (5.2) with (4.23), (4.29) and (4.34) we obtain:

$$-\frac{2}{(2\pi)^{3/2}}v^t(\mathbf{p}-\mathbf{p}')$$

$$= \frac{(k_\mu^2+p^2)(k_\mu^2+p'^2)}{2k_\mu^2} \sum_{anlm} \left(\frac{Z_a}{k_\mu n}\right) e^{-i(\mathbf{p}-\mathbf{p}')\cdot\mathbf{X}_a} \chi_{n,l,m}^t(\mathbf{p})\chi_{n,l,m}^{t*}(\mathbf{p}')$$

$$(5.3)$$

where

$$\chi_{n,l,m}^t(\mathbf{p}) = M(p)Y_{n-1,l,m}(\mathbf{u}) \tag{5.4}$$

If we let

$$\xi_\tau^t(\mathbf{p}) \equiv \sqrt{\frac{Z_a}{k_\mu n}}\; e^{-i\mathbf{p}\cdot\mathbf{X}_a}\chi_{n,l,m}^t(\mathbf{p}) \tag{5.5}$$

where τ stands for the indices $\{a,n,l,m\}$, then we can write equation (5.3) in the form:

$$-\frac{2}{(2\pi)^{3/2}}v^t(\mathbf{p}-\mathbf{p}') \;\; = \;\; \frac{(k_\mu^2+p^2)(k_\mu^2+p'^2)}{2k_\mu^2} \sum_{\tau'} \xi_{\tau'}^t(\mathbf{p})\xi_{\tau'}^{*t}(\mathbf{p}') \tag{5.6}$$

With the help of (5.6), the momentum-space Schrödinger equation

$$(k_\mu^2+p^2)\varphi_\mu^t(\mathbf{p}) = -\frac{2}{(2\pi)^{3/2}}\int d^3p'\; v^t(\mathbf{p}-\mathbf{p}')\varphi_\mu^t(\mathbf{p}') \tag{5.7}$$

can be rewritten in the form

$$\varphi_\mu^t(\mathbf{p}) = \sum_{\tau'} \xi_{\tau'}^t(\mathbf{p}) \int d^3p'\; \left(\frac{k_\mu^2+p'^2}{2k_\mu^2}\right) \xi_{\tau'}^{*t}(\mathbf{p}')\varphi_\mu^t(\mathbf{p}') \tag{5.8}$$

Then if we let

$$\varphi_\mu^t(\mathbf{p}) = \sum_\tau \xi_\tau^t(\mathbf{p})C_{\tau,\mu} \tag{5.9}$$

equation (5.8) becomes

$$k_\mu \sum_{\tau'} \xi_{\tau'}^t(\mathbf{p})C_{\tau',\mu} = \sum_{\tau'} \xi_{\tau'}^t(\mathbf{p}) \sum_\tau K_{\tau',\tau}C_{\tau,\mu} \tag{5.10}$$

where

$$K_{\tau',\tau} \equiv \int d^3p \left(\frac{k_\mu^2 + p^2}{2k_\mu}\right) \xi_{\tau'}^{*t}(\mathbf{p})\xi_\tau^t(\mathbf{p})$$

$$= \sqrt{\frac{Z_{a'}Z_a}{n'n}} \int d^3p \left(\frac{k_\mu^2 + p^2}{2k_\mu^2}\right) e^{i\mathbf{p}\cdot(\mathbf{X}_{a'}-\mathbf{X}_a)}\chi_{n',l'm'}^{t*}(\mathbf{p})\chi_{n,l,m}^t(\mathbf{p})$$

$$(5.11)$$

Equation (5.10) will be satisfied provided that

$$\sum_\tau [K_{\tau',\tau} - k_\mu\delta_{\tau',\tau}] C_{\tau,\mu} = 0 \qquad (5.12)$$

The matrix $K_{\tau',\tau}$ can be rewritten in the form [263, 217, 222, 223]

$$K_{\tau',\tau} = \sqrt{\frac{Z_{a'}Z_a}{n'n}} \int d\Omega \; e^{i\mathbf{p}\cdot(\mathbf{X}_{a'}-\mathbf{X}_a)}Y_{n'-1,l',m'}^*(\mathbf{u})Y_{n-1,l,m}(\mathbf{u}) \qquad (5.13)$$

or

$$K_{\tau',\tau} = \sqrt{\frac{Z_{a'}Z_a}{n'n}} \; S_{n'l'm'}^{n,l,m}(\mathbf{X}_{a'} - \mathbf{X}_a) \qquad (5.14)$$

where

$$S_{n'l'm'}^{n,l,m}(\mathbf{R}) \equiv \int d\Omega \; e^{i\mathbf{p}\cdot\mathbf{R}}Y_{n'-1,l',m'}^*(\mathbf{u})Y_{n-1,l,m}(\mathbf{u}) \qquad (5.15)$$

We shall call the hyperangular integrals $S_{n'l'm'}^{n,l,m}$ *Shibuya-Wulfman integrals* to honor these two pioneers of momentum-space quantum theory.

As a simple example to illustrate (5.12), we can think of an electron moving in the Coulomb potential of two nuclei, with nuclear charges Z_1 and Z_2, located respectively at positions $\mathbf{X}_1$ and $\mathbf{X}_2$. In the crude approximation where we use only a single $1s$ orbital on each nucleus, we can represent the electronic wave function of this system by:

$$\varphi_\mu^t(\mathbf{p}) \approx \xi_1^t(\mathbf{p})C_{1,\mu} + \xi_2^t(\mathbf{p})C_{2,\mu} \qquad (5.16)$$

where

$$\xi_1^t(\mathbf{p}) \equiv \sqrt{Z_1} \; M(p)e^{-i\mathbf{p}\cdot\mathbf{X}_1}Y_{000}(\mathbf{u}) = \frac{1}{\pi}\sqrt{\frac{Z_1}{2}}M(p) \; e^{-i\mathbf{p}\cdot\mathbf{X}_1}$$

$$\xi_2^t(\mathbf{p}) \equiv \sqrt{Z_2} \; M(p)e^{-i\mathbf{p}\cdot\mathbf{X}_2}Y_{000}(\mathbf{u}) = \frac{1}{\pi}\sqrt{\frac{Z_2}{2}}M(p) \; e^{-i\mathbf{p}\cdot\mathbf{X}_2} \qquad (5.17)$$

Using methods which will be discussed in a later section, we can evaluate the Shibuya-Wulfman integrals. The result is:

$$K_{\tau',\tau} = \begin{pmatrix} Z_1 & \sqrt{Z_1 Z_2}\, e^{-s}(1+s) \\ \sqrt{Z_1 Z_2}\, e^{-s}(1+s) & Z_2 \end{pmatrix} \tag{5.18}$$

where

$$s \equiv k_\mu R = k_\mu |\mathbf{X}_1 - \mathbf{X}_2| \tag{5.19}$$

The secular equation, (5.12), then requires that

$$\begin{vmatrix} Z_1 - k_\mu & \sqrt{Z_1 Z_2}\, e^{-s}(1+s) \\ \sqrt{Z_1 Z_2}\, e^{-s}(1+s) & Z_2 - k_\mu \end{vmatrix} = 0 \tag{5.20}$$

or

$$2k_\mu = Z_1 + Z_2 \pm \sqrt{(Z_1 + Z_2)^2 + 4Z_1 Z_2 \left[(1+s)^2 e^{-2s} - 1\right]} \tag{5.21}$$

We can use equation (5.21) to generate a curve representing the ground-state energy of the system as a function of the internuclear distance, R. We do this by picking a value of s, substituting it into (5.21) to find the corresponding value of k_μ, and then finding the internuclear separation and energies by means of the relationships, $R = s/k_\mu$ and $\epsilon = -k_\mu^2/2$. In the united-atom limit, $R = 0$, the $+$-root of the quadratic equation, (5.21), yields the exact ground-state energy of the system.

$$p_{0+} = Z_1 + Z_2 \qquad \epsilon_+ = -\frac{(Z_1 + Z_2)^2}{2} \tag{5.22}$$

In the separated-atom limit, $R = \infty$, equation (5.21) also yields exact energies:

$$p_{0+} = Z_1 \qquad \epsilon_+ = -\frac{Z_1^2}{2}$$
$$p_{0-} = Z_2 \qquad \epsilon_- = -\frac{Z_2^2}{2} \tag{5.23}$$

For intermediate values of R, the ground-state energies predicted by (5.21) are appreciably above the best available values found from direct space calculations; but when more basis functions are added, the

Sturmian expansion converges rapidly towards the best available wave functions and energies, as illustrated in Figure 6.1. By using many basis functions in a calculation of the type sketched above, but with a secular equation based on a second-iterated form of the wave equation, Koga and Matsuhashi [190-192] obtained electronic energies with 10-figure accuracy for the H_2^+ ion. Their energies at various internuclear separations agreed with the best available values from direct-space calculations; but the Koga-Matsuhashi values were of higher precision.

The simple example discussed above illustrates several interesting features of Sturmian basis sets: In quantum theory, we are accustomed to solving secular equations in which a Hamiltonian matrix is diagonalized, yielding roots which correspond to the energies of the system; and the basis functions are entirely determined before we begin to construct a matrix representation of the Hamiltonian. However, the Sturmian secular equation, (5.12), is of a completely different form: The roots are not energies, but k_μ values, related to the one-electron energies by $\epsilon = -k_\mu^2/2$. These k_μ values determine the exponential decay of the basis functions in direct-space. The rapid convergence of Sturmian expansions is due to the fact that all the basis functions used to represent a state of the system have the correct exponential decay. The values of k_μ for each state are not known before the secular equations are solved; and hence the basis functions are not known in advance - the form is known, but not the scale. By solving the Sturmian secular equations, we obtain in one stroke, not only the spectrum of energies, but also the optimal basis set for each state. However, this double victory is won at a price: We do not know the size of the system in direct space until we have finished the calculation! In the simple example discussed above, we could not specify the internuclear separation in advance; but of course it was easy to generate the curve $\epsilon(R)$ shown in Figure 6.1 by leting the parameter s run over a range of values. The fact that even the crude approximation of equations (5.16) and (5.17) yielded exact energies in the united-atom and separated-atom limits is due to the automatic basis-set optimization which is built into the method. The basis set adjusted automatically to the enormous change in internuclear distance, from $R = 0$ to $R = \infty$.

Combined Coulomb attraction and repulsion

Since equation (5.3) can be derived entirely from the properties of the Gegenbauer polynomials and the 4-dimensional hyperspherical harmonics, it must also hold if some of the Z_a's are negative. Thus equation (5.3) may also be used to treat the case of an electron moving in the potential of a many-center charge distribution where some of the charges are negative, but where the sum of the charges is positive, so that bound states exist. If we let γ_a be defined by

$$Z_a = \gamma_a |Z_a| \qquad (5.24)$$

where $\gamma_a = \pm 1$, and if we let

$$\xi_\tau^t(\mathbf{p}) \equiv \sqrt{\frac{|Z_a|}{k_\mu n}} \; e^{i\mathbf{p}\cdot\mathbf{X}_a} \chi_{n,l,m}^t(\mathbf{p}) \qquad (5.25)$$

where τ stands for the set of indices $\{a, n, l, m\}$, then (5.3) becomes:

$$-\frac{2}{(2\pi)^{3/2}} v^t(\mathbf{p} - \mathbf{p}') = \frac{(k_\mu^2 + p^2)(k_\mu^2 + p'^2)}{2k_\mu^2} \sum_\tau \gamma_\tau \xi_\tau^t(\mathbf{p}) \xi_\tau^{t*}(\mathbf{p}') \qquad (5.26)$$

and the reciprocal-space Schrödinger equation can be written in the form:

$$\varphi_\mu^t(\mathbf{p}) = \sum_\tau \gamma_\tau \xi_\tau^t(\mathbf{p}) \int d^3p' \left(\frac{k_\mu^2 + p'^2}{2k_\mu^2}\right) \xi_\tau^{t*}(\mathbf{p}') \varphi_\mu^t(\mathbf{p}') \qquad (5.27)$$

The Schrödinger equation will have a solution of the form $\varphi_\mu^t(\mathbf{p}) = \sum_\tau \xi_\tau^t(\mathbf{p}) C_{\tau,\mu}$ if k_μ and the coefficients $C_{\tau,\mu}$ satisfy the set of secular equations

$$\sum_\tau \left[\gamma_{\tau'} K_{\tau',\tau} - k_\mu \delta_{\tau',\tau} \right] C_{\tau,\mu} = 0 \qquad (5.28)$$

where $K_{\tau',\tau}$ is defined by equations (5.11), and (5.13).

To illustrate combined Coulomb attraction and repulsion, we can repeat the simple example shown in equations (5.16)-(5.23) for the case where Z_1 is positive, while Z_2 is negative and $|Z_1| > |Z_2|$. Then $\gamma_1 = 1$

and $\gamma_2 = -1$; and if we limit the basis set to a single $1s$ orbital on each center (as before), the secular equations require that

$$\begin{vmatrix} |Z_1| - k_\mu & \sqrt{|Z_1||Z_2|}\, e^{-s}(1+s) \\ -\sqrt{|Z_1||Z_2|}\, e^{-s}(1+s) & -|Z_2| - k_\mu \end{vmatrix} = 0 \qquad (5.29)$$

or

$$2k_\mu = |Z_1| - |Z_2| \pm \sqrt{(|Z_1| - |Z_2|)^2 - 4|Z_1||Z_2|\left[(1+s)^2 e^{-2s} - 1\right]} \quad (5.30)$$

For $s = 0$ we obtain the exact ground-state energy of the system:

$$p_{0+} = |Z_1| - |Z_2| \qquad \epsilon_+ = -\frac{(|Z_1| - |Z_2|)^2}{2} \qquad (5.31)$$

and in the limit $s = \infty$ we again obtain exact solutions, but only the positive root of the quadratic equation for k_μ corresponds to a bound state:

$$p_{0+} = |Z_1| \qquad \epsilon_+ = -\frac{|Z_1|^2}{2}$$
$$p_{0-} = -|Z_2| \quad \text{(not bound)} \qquad (5.32)$$

Shibuya-Wulfman integrals

From equation (5.13), we can see that the problem of solving the Schrödinger equation for a particle moving in a many-center Coulomb potential can be reduced to the problem of calculating integrals of the form [272, 170]:

$$S_{n'l'm'}^{nlm}(\mathbf{R}) = \int d\Omega\; e^{i\mathbf{p}\cdot\mathbf{R}}\, Y_{n'-1,l',m'}^*(\mathbf{u}) Y_{n-1,l,m}(\mathbf{u}) \qquad (5.33)$$

after which the matrix of hyperangular overlap integrals can be diagonalized using standard programs. The basic information needed for the evaluation of integrals of this type is given to us by Fock's relationship:

$$\chi_{n,l,m}^t(\mathbf{p}) = M(p) Y_{n-1,l,m}(\mathbf{u}) \qquad (5.34)$$

from which it follows that

$$\frac{1}{(2\pi)^{3/2}} \int d^3p \; e^{i\mathbf{p}\cdot\mathbf{R}} \; M(p) Y_{n-1,l,m}(\mathbf{u}) = \chi_{nlm}(\mathbf{R}) \qquad (5.35)$$

From (2.14) and (4.21) we have:

$$d^3p \; M(p) = d\Omega \; \frac{k_\mu^2 + p^2}{2\sqrt{k_\mu}} \qquad (5.36)$$

and from (2.13),

$$1 + u_4 = \frac{2k_\mu^2}{k_\mu^2 + p^2} \qquad (5.37)$$

so that (5.35) becomes:

$$\left(\frac{k_\mu}{2\pi}\right)^{3/2} \int d\Omega \; \frac{e^{i\mathbf{p}\cdot\mathbf{R}}}{1 + u_4} Y_{n-1,l,m}(\mathbf{u}) = \chi_{nlm}(\mathbf{R}) \qquad (5.38)$$

By comparing (5.33) with (5.38), we can see that by finding the coefficients in the series:

$$(1 + u_4) Y_{\mu'}^*(\mathbf{u}) Y_\mu(\mathbf{u}) = \sum_{\mu''} Y_{\mu''}(\mathbf{u}) c_{\mu',\mu}^{\mu''} \qquad (5.39)$$

we can write the hyperangular overlap integrals as a series of hydrogenlike orbitals whose argument is the internuclear distance vector $\mathbf{R}$:

$$\int d\Omega e^{i\mathbf{p}\cdot\mathbf{R}} \; Y_{\mu'}^*(\mathbf{u}) Y_\mu(\mathbf{u}) = \left(\frac{2\pi}{k_\mu}\right)^{3/2} \sum_{\mu''} \chi_{\mu''}(\mathbf{R}) c_{\mu',\mu}^{\mu''} \qquad (5.40)$$

where we have let μ stand for the set of indices $\{n, l, m\}$. From the orthonormality of the hyperspherical harmonics, it follows that the coefficients are given by:

$$c_{\mu',\mu}^{\mu''} = \int d\Omega \; (1 + u_4) Y_{\mu''}^*(\mathbf{u}) Y_{\mu'}^*(\mathbf{u}) Y_\mu(\mathbf{u}) \qquad (5.41)$$

Shibuya and Wulfman used the Wigner coefficients of the hyperspherical harmonics to evaluate $c_{\mu',\mu}^{\mu''}$. Alternatively, we can use the angular integration theorem shown in equation (3.49), which, in this application, becomes:

$$\int d\Omega \prod_{j=1}^{4} u_j^{n_j} = \begin{cases} \dfrac{4\pi^2}{(n+2)!!} \prod_{j=1}^{4}(n_j - 1)!! & \text{all } n_j's \text{ even} \\ 0 & \text{otherwise} \end{cases} \qquad (5.42)$$

For example, suppose that we wish to evaluate the hyperangular overlap integral:

$$\int d\Omega \; e^{i\mathbf{p}\cdot\mathbf{R}} \; Y_{000}^{*}(\mathbf{u})Y_{000}(\mathbf{u}) = \frac{1}{2\pi^2} \int d\Omega e^{i\mathbf{p}\cdot\mathbf{R}} \qquad (5.43)$$

Then, from (5.41), (5.42) and Table 2.3, we obtain

$$c_{\{000\},\{000\}}^{\{000\}} = \frac{1}{(\sqrt{2\pi})^3} \int d\Omega \; (1 + u_4) = \frac{1}{\sqrt{2\pi}}$$

$$c_{\{000\},\{000\}}^{\{100\}} = \frac{-1}{(\sqrt{2\pi})^3} \int d\Omega \; (1 + u_4)2u_4 = \frac{-1}{2\sqrt{2\pi}} \qquad (5.44)$$

all other coefficients being zero. These coefficients can now be substituted into equation (5.40); and, with the help of Table 2.2, we obtain the result which we used in the previous section:

$$\frac{1}{2\pi^2} \int d\Omega e^{i\mathbf{p}\cdot\mathbf{R}} = e^{-s}(1 + s) \qquad (5.45)$$

A third method [38] for evaluating the hyperangular overlap integrals begins (like the other two methods) with the relationship shown in equation (5.35). We then use (1.8) and (2.16) to write down explicit expressions for $Y_{n-1,l,m}(\mathbf{u})$ and $\chi_{n,l,m}(\mathbf{R})$. This gives us the equation:

$$\frac{1}{2\pi^2} \int d\Omega \; e^{i\mathbf{p}\cdot\mathbf{R}} \frac{1}{1 + u_4} C_{n-l-1}^{l+1}(u_4) \sin^l \chi Y_{lm}(\theta_p, \phi_p) \qquad (5.46)$$

$$= \frac{(-1)^{n-l-1}2i^l(n+l)!e^{-s}F(l+1-n|2l+2|2s)}{(2l+1)!l!n(n-l-1)!}s^l Y_{lm}(\theta_s, \phi_s)$$

where $s \equiv k_\mu R$. If we let

$$s_j \equiv k_\mu R_j \qquad (5.47)$$

then we can define the lth-order harmonic polynomial, h_l, by the relationship:

$$s^l Y_{lm}(\theta_s, \phi_s) \equiv h_l(s_1, s_2, s_3) \qquad (5.48)$$

Similarly, using (4.18) and (4.19), we can write:

$$\sin^l \chi Y_{lm}(\theta_p, \phi_p) \equiv h_l(u_1, u_2, u_3) \tag{5.49}$$

where $h_l(u_1, u_2, u_3)$ is the same harmonic polynomial which appears in (5.48), but with the s_j's replaced by u_j's. Thus (5.46) becomes:

$$\frac{1}{2\pi^2} \int d\Omega\ e^{i\mathbf{p}\cdot\mathbf{R}}\ \frac{1}{1+u_4} C_{n-l-1}^{l+1}(u_4) h_l(u_j)$$
$$= \frac{(-1)^{n-l-1} 2i^l (n+l)! e^{-s} F(l+1-n|2l+2|2s) h_l(s_j)}{(2l+1)!\,l!\,n(n-l-1)!} \tag{5.50}$$

where we have also assumed the harmonic polynomials to be real. Equation (5.50) is still not very convenient to use, because of the Gegenbauer polynomial on the left-hand side, and because of the awkward factor $1 + u_4$ in the denominator. Both difficulties can be removed by using the relationship [162],

$$u_4^k = \frac{l!k!}{2^k} \sum_{q=0}^{[k/2]} \frac{(k+l+1-2q)}{q!(k+l+1-q)!} C_{k-2q}^{l+1}(u_4) \tag{5.51}$$

from which we have:

$$\frac{1}{2\pi^2} \int d\Omega\ e^{i\mathbf{p}\cdot\mathbf{R}}\ \frac{1}{1+u_4} u_4^k h_l(u_j)$$
$$= \frac{(-1)^k i^l k! e^{-s} h_l(s_j)}{2^k (2l+1)!} \sum_{q=0}^{[k/2]} \frac{2(k+2l+1-2q)! F(2q-k|2l+2|2s)}{q!(k+l+1-q)!(k-2q)!} \tag{5.52}$$

A similar relation, of course, exists for $k + 1$; and if we add this to (5.52), we can cancel out the factor $1 + u_4$. Thus we obtain the useful relationship:

$$\frac{1}{2\pi^2} \int d\Omega\ e^{i\mathbf{p}\cdot\mathbf{R}}\ u_4^k h_l(u_j) = \mathcal{F}_{k,l}(s) h_l(s_j) \tag{5.53}$$

where $\mathcal{F}_{k,l}(s)$ is the easily-programmed function

$$\mathcal{F}_{k,l}(s) \equiv \frac{(-1)^k i^l k! e^{-s}}{2^k (2l+1)!} \left[\sum_{q=0}^{[k/2]} \frac{2(k+2l+1-2q)! F(2q-k|2l+2|2s)}{q!(k+l+1-q)!(k-2q)!} \right.$$

$$-(k+1)\sum_{q=0}^{[(k+1)/2]}\frac{(k+2l+2-2q)!F(2q-k-1|2l+2|2s)}{q!(k+l+2-q)!(k+1-2q)!}\Bigg]$$

(5.54)

For example, when $h_l = 1$, (5.53) becomes:

$$\frac{1}{2\pi^2}\int d\Omega e^{i\mathbf{p}\cdot\mathbf{R}} = \mathcal{F}_{0,0}(s)$$

(5.55)

With a little work, one can show that

$$\mathcal{F}_{0,0}(s) = e^{-s}(1+s)$$

(5.56)

which checks with equation (5.45). Hyperangular overlap integrals derived from equation (5.53) are shown in Table 5.1 for the first few atomic orbitals. In most cases, the derivation of these integrals is straightforward; but the integrals linking $2p_j$ with $2p_j$ on another center require a few additional words of explanation: From Table 2.4 we have (for example):

$$\int d\Omega e^{i\mathbf{p}\cdot\mathbf{R}}\, Y^*_{2p_1}(\mathbf{u})Y_{2p_1}(\mathbf{u}) = \frac{4}{2\pi^2}\int d\Omega e^{i\mathbf{p}\cdot\mathbf{R}}u_1^2$$

(5.57)

Comparing this with equation (5.53), we can see that we are dealing with a case where $k = 0$, since u_4 does not appear. However, to apply (5.53), we have to resolve u_1^2 into a series of polynomials which are harmonic in u_1, u_2 and u_3. This can be done with the help of Table 3.1; but we must remember that (from equation (4.16))

$$u_1^2 + u_2^2 + u_3^2 = 1 - u_4^2$$

(5.58)

Thus the correct resolution of u_1^2 into orbital angular momentum eigenfunctions is given by

$$u_1^2 = h_3 + (1 - u_4^2)h_0 = \left[u_1^2 - \frac{1}{3}(u_1^2 + u_2^2 + u_3^2)\right] + \frac{1}{3}(1 - u_4^2)$$

(5.59)

and (5.53) yields the result:

$$\int d\Omega e^{i\mathbf{p}\cdot\mathbf{R}}\, Y^*_{2p_1}(\mathbf{u})Y_{2p_1}(\mathbf{u}) = 4\left[\mathcal{F}_{0,2}(s)\left(s_1^2 - \frac{1}{3}s^2\right) + \frac{1}{3}\mathcal{F}_{0,0}(s) - \frac{1}{3}\mathcal{F}_{2,0}(s)\right]$$

(5.60)

Shibuya-Wulfman integrals and translations

One can think of the Shibuya-Wulfman integrals as representations of the translation group based on one-particle hydrogenlike Sturmians. We can understand this interpretation of the Shibuya-Wulfman integrals by means of the following argument: From equation (2.1) we have the relationship

$$\chi_{n,l,m}(\mathbf{x}) = \frac{1}{(2\pi)^{3/2}} \int d^3p \; e^{i\mathbf{p}\cdot\mathbf{x}} \chi_{n,l,m}^t(\mathbf{p}) \tag{5.61}$$

so that

$$\chi_{n,l,m}(\mathbf{x} + \mathbf{R}) = \frac{1}{(2\pi)^{3/2}} \int d^3p \; e^{i\mathbf{p}\cdot(\mathbf{x}+\mathbf{R})} \chi_{n,l,m}^t(\mathbf{p}) \tag{5.62}$$

We can make a Sturmian expansion of a 3-dimensional plane wave analogous to the d-dimensional expansion shown in equations (4.58)-(4.60). We let

$$e^{i\mathbf{p}\cdot\mathbf{x}} = \left(\frac{k_\mu^2 + p^2}{2k_\mu^2}\right) \sum_{nlm} \chi_{n,l,m}^{t*}(\mathbf{p}) a_{nlm} \tag{5.63}$$

where a_{nlm} are coefficients which are independent of $\mathbf{p}$, and which we must determine. To find these coefficients, we multiply both sides of (5.63) by $\chi_{n',l',m'}^t(\mathbf{p})$ and integrate over d^3p. This gives us

$$\int d^3p \; \chi_{n,l,m}^t(\mathbf{p}) e^{i\mathbf{p}\cdot\mathbf{x}} = \sum_{nlm} a_{nlm} \int d^3p \left(\frac{k_\mu^2 + p^2}{2k_\mu^2}\right) \chi_{n,l,m}^{t*}(\mathbf{p}) \chi_{n',l',m'}^t(\mathbf{p}) \tag{5.64}$$

Then, making use of the momentum-space orthonormality relations, (4.44), and equation (5.61), we have

$$a_{nlm} = (2\pi)^{3/2} \chi_{n,l,m}(\mathbf{x}) \tag{5.65}$$

so that

$$e^{i\mathbf{p}\cdot\mathbf{x}} = (2\pi)^{3/2} \left(\frac{k_\mu^2 + p^2}{2k_\mu^2}\right) \sum_{nlm} \chi_{n,l,m}^{t*}(\mathbf{p}) \chi_{n,l,m}(\mathbf{x}) \tag{5.66}$$

If we substitute this expansion into (5.62), we obtain

$$\chi_{n,l,m}(\mathbf{x}+\mathbf{R}) = \sum_{n'l'm'} \chi_{n',l',m'}(\mathbf{x}) \int d^3p \left(\frac{k_\mu^2 + p^2}{2k_\mu^2} \right) e^{i\mathbf{p}\cdot\mathbf{R}} \chi_{n',l',m'}^{t*}(\mathbf{p}) \chi_{n,l,m}^t(\mathbf{p})$$

(5.67)

Since

$$\chi_{n,l,m}^t(\mathbf{p}) = M(p) Y_{n-1,l,m}(\mathbf{u})$$

(5.68)

where

$$M(p) = \frac{k_\mu^{5/2}}{(k_\mu^2 + p^2)^2}$$

(5.69)

and since

$$M^2(p) \left(\frac{k_\mu^2 + p^2}{2k_\mu^2} \right) = \left(\frac{2k_\mu}{k_\mu^2 + p^2} \right)^3$$

(5.70)

we can identify the momentum-space integral in (5.67) as a Shibuya-Wulfman hyperangular integral:

$$\int d^3p \left(\frac{k_\mu^2 + p^2}{2k_\mu^2} \right) e^{i\mathbf{p}\cdot\mathbf{R}} \chi_{n',l',m'}^{t*}(\mathbf{p}) \chi_{n,l,m}^t(\mathbf{p})$$

$$= \int d\Omega \, e^{i\mathbf{p}\cdot\mathbf{R}} Y_{n'-1,l',m'}^*(\mathbf{u}) Y_{n-1,l,m}(\mathbf{u}) \equiv S_{n'l'm'}^{nlm}(\mathbf{R})$$

(5.71)

Substituting (5.71) into (5.67), we have

$$\chi_{n,l,m}(\mathbf{x}+\mathbf{R}) = \sum_{n'l'm'} \chi_{n',l',m'}(\mathbf{x}) S_{n'l'm'}^{nlm}(\mathbf{R})$$

(5.72)

from which it follows that

$$\sum_{n''l''m''} S_{n'l'm'}^{n''l''m''}(\mathbf{R}) S_{n''l''m''}^{nlm}(\mathbf{R}') = S_{n'l'm'}^{nlm}(\mathbf{R}+\mathbf{R}')$$

(5.73)

Thus the Shibuya-Wulfman integrals can be thought of as elements of a continuous group representing translations. If we multiply equation (5.72) on both sides by $\chi_{n'',l'',m''}^*(\mathbf{x})/r$ and integrate over d^3x, making use of the potential-weighted orthonormality relation, (4.53), we obtain the useful relationship:

$$S_{n''l''m''}^{nlm}(\mathbf{R}) = \frac{n}{k_\mu} \int d^3x \, \chi_{n'',l'',m''}^*(\mathbf{x}) \frac{1}{r} \chi_{n,l,m}(\mathbf{x}+\mathbf{R})$$

(5.74)

From (5.74) we can see that the Shibuya-Wulfman integrals may also be interpreted as nuclear attraction integrals involving one-electron hydrogenlike Sturmians.

Table 5.1: Shibuya-Wulfman integrals

α	α'	$\int d\Omega \; e^{i\mathbf{p}\cdot\mathbf{R}} Y_\alpha^*(\mathbf{u}) Y_{\alpha'}(\mathbf{u})$
$1s$	$1s$	$\mathcal{F}_{0,0}(s)$
$1s$	$2p_j$	$-2i\mathcal{F}_{0,1}(s)s_j \qquad j = 1,2,3$
$1s$	$2s$	$-2\mathcal{F}_{1,0}(s)$
$2p_j$	$1s$	$2i\mathcal{F}_{0,1}(s)s_j \qquad j = 1,2,3$
$2p_j$	$2p_j$	$4\mathcal{F}_{0,2}(s)\left[s_j^2 - \dfrac{1}{3}s^2\right] + \dfrac{4}{3}\left[\mathcal{F}_{0,0}(s) - \mathcal{F}_{2,0}(s)\right]$
$2p_j$	$2p_k$	$4\mathcal{F}_{0,2}(s)s_j s_k \qquad j \neq k$
$2p_j$	$2s$	$-4i\mathcal{F}_{1,1}(s)s_j$
$2s$	$2s$	$4\mathcal{F}_{2,0}(s)$

Table 5.1 continued

α	$\int d\Omega \, e^{i\mathbf{p}\cdot\mathbf{R}} Y_{1s}^*(\mathbf{u}) Y_\alpha(\mathbf{u})$	$\int d\Omega \, e^{i\mathbf{p}\cdot\mathbf{R}} Y_{2s}^*(\mathbf{u}) Y_\alpha(\mathbf{u})$
$3s$	$4\mathcal{F}_{2,0}(s) - \mathcal{F}_{0,0}(s)$	$-8\mathcal{F}_{3,0}(s) + 2\mathcal{F}_{1,0}(s)$
$3p_j$	$2i\sqrt{6}\mathcal{F}_{1,1}(s)s_j$	$-4i\sqrt{6}\mathcal{F}_{2,1}(s)s_j$
$3d_{xy}$	$-2\sqrt{6}\mathcal{F}_{0,2}(s)s_1 s_2$	$4\sqrt{6}\mathcal{F}_{1,2}(s)s_1 s_2$
$3d_{xz}$	$-2\sqrt{6}\mathcal{F}_{0,2}(s)s_1 s_3$	$4\sqrt{6}\mathcal{F}_{1,2}(s)s_1 s_3$
$3d_{yz}$	$-2\sqrt{6}\mathcal{F}_{0,2}(s)s_2 s_3$	$4\sqrt{6}\mathcal{F}_{1,2}(s)s_2 s_3$
$3d_{x^2-y^2}$	$-\sqrt{6}\mathcal{F}_{0,2}(s)(s_1^2 - s_2^2)$	$2\sqrt{6}\mathcal{F}_{1,2}(s)(s_1^2 - s_2^2)$
$3d_{z^2}$	$-\sqrt{2}\mathcal{F}_{0,2}(s)(2s_3^2 - s_1^2 - s_2^2)$	$2\sqrt{2}\mathcal{F}_{1,2}(s)(2s_3^2 - s_1^2 - s_2^2)$

Table 5.1 continued

α	$\int d\Omega \; e^{i\mathbf{p}\cdot\mathbf{R}} Y_{3s}^*(\mathbf{u}) Y_\alpha(\mathbf{u})$
$3s$	$16\mathcal{F}_{4,0}(s) - 8\mathcal{F}_{2,0}(s) + \mathcal{F}_{0,0}(s)$
$2p_j$	$-2i\left[4\mathcal{F}_{2,1}(s) - \mathcal{F}_{0,1}(s)\right] s_j$
$3p_j$	$2i\sqrt{6}\left[4\mathcal{F}_{3,1}(s) - \mathcal{F}_{1,1}(s)\right] s_j$
$3d_{xy}$	$-2\sqrt{6}\left[4\mathcal{F}_{2,2}(s) - \mathcal{F}_{0,2}(s)\right] s_1 s_2$
$3d_{z^2}$	$-\sqrt{2}\left[4\mathcal{F}_{2,2}(s) - \mathcal{F}_{0,2}(s)\right]\left(2s_3^2 - s_1^2 - s_2^2\right)$
$3d_{x^2-y^2}$	$-\sqrt{6}\left[4\mathcal{F}_{2,2}(s) - \mathcal{F}_{0,2}(s)\right]\left(s_1^2 - s_2^2\right)$

Table 5.2 (see equation (5.54))

k	l	$\mathcal{F}_{k,l}(s) \qquad s \equiv k_\mu R$
0	0	$(1+s)e^{-s}$
0	1	$\dfrac{i}{3}(1+s)e^{-s}$
0	2	$-\dfrac{1}{12}(1+s)e^{-s}$
1	0	$\dfrac{1}{3}s^2 e^{-s}$
1	1	$-\dfrac{i}{12}(1+s-s^2)e^{-s}$
1	2	$-\dfrac{1}{60}(-2-2s+s^2)e^{-s}$

Table 5.2 continued

k	l	$\mathcal{F}_{k,l}(s)$ $\qquad s \equiv k_\mu R$
2	0	$\dfrac{1}{12}(3 + 3s - 2s^2 + s^3)e^{-s}$
2	1	$\dfrac{i}{60}(5 + 5s - 4s^2 + s^3)e^{-s}$
2	2	$\dfrac{1}{360}(-9 - 9s + 6s^2 - s^3)e^{-s}$
3	0	$\dfrac{s^2}{60}(15 - 5s + s^2)e^{-s}$
3	1	$-\dfrac{i}{360}(15 + 15s - 27s^2 + 8s^3 - s^4)e^{-s}$
3	2	$\dfrac{1}{2520}(42 + 42s - 45s^2 + 11s^3 - s^4)e^{-s}$

Table 5.3: Shibuya-Wulfman integrals, $S^{\alpha}_{\alpha'}$

α'	$\alpha = 1s$	$\alpha = 2s$
$1s$	$(1+s)e^{-s}$	$-2s^2e^{-s}/3$
$2s$	$-2s^2e^{-s}/3$	$(3+3s-2s^2+s^3)e^{-s}/3$
$2p_j$	$2s_j(1+s)e^{-s}/3$	$s_j(1+s-s^2)e^{-s}/3$

α'	$\alpha = 2p_1$
$2p_1$	$(3+3s+s^2-s_1^2-ss_1^2)e^{-s}/3$
$2p_2$	$s_1s_2(1+s)e^{-s}/3$
$2p_3$	$s_1s_3(1+s)e^{-s}/3$

Exercises

1. Calculate the Shibuya-Wulfman integrals $S_{1,0,0}^{1,0,0}$, $S_{1,0,0}^{1,0,0}$, and $S_{1,0,0}^{2,1,0}$ in terms of the universal function $\mathcal{F}_{k,l}(s)$ by means of equation (5.53). Compare your results with those shown in Table 5.1. Is the integral $S_{1,0,0}^{2,1,0}$ real?

2. Calculate
$$S_{1,0,0}^{2,1,0} = \int d\Omega \; e^{i\mathbf{p}\cdot\mathbf{R}} Y_{0,0,0}^*(\mathbf{u}) Y_{1,1,0}(\mathbf{u})$$
by means of equations (5.40)-(5.42). Show that the result agrees with Table 5.1 and Exercise 1.

3. Use (3.49) to write down an angular integration theorem analogous to (5.42) for the case where $d = 3$.

Chapter 6

ITERATION OF THE WAVE EQUATION

Integral forms of the many-particle wave equation

In Chapter 4, we saw that the momentum-space Schrödinger equation for a many-particle system has the form:

$$(p_0^2 + p^2)\psi^t(\mathbf{p}) = -\frac{2}{(2\pi)^{d/2}} \int dp' \, V^t(\mathbf{p} - \mathbf{p}')\psi^t(\mathbf{p}') \qquad (6.1)$$

In equation (6.1), $d = 3N$, where N is the number of particles in the system. According to the Fourier convolution theorem, the Fourier transform of the product of two functions is the scalar product of their Fourier transforms, and this theorem has the d-dimensional generalization:

$$\int dx \, e^{-i\mathbf{p}\cdot\mathbf{x}} f(\mathbf{x})g(\mathbf{x}) = \int dp' f^t(\mathbf{p} - \mathbf{p}')g^t(\mathbf{p}') \qquad (6.2)$$

where $e^{-i\mathbf{p}\cdot\mathbf{x}}$ is a d-dimensional plane wave. If we compare (6.2) with (6.1), letting $f(\mathbf{x}) = V(\mathbf{x})$ and $g(\mathbf{x}) = \psi(\mathbf{x})$, we can see that it is possible to write the many-particle Schrödinger equation in the form:

$$(p_0^2 + p^2)\psi^t(\mathbf{p}) = -\frac{2}{(2\pi)^{d/2}} \int dx \, e^{-i\mathbf{p}\cdot\mathbf{x}} V(\mathbf{x})\psi(\mathbf{x}) \qquad (6.3)$$

Dividing both sides of (6.3) by $p_0^2 + p^2$ and taking the Fourier transform, we obtain another form of the Schrödinger equation:

$$\psi(\mathbf{x}) = -\int d\mathbf{x}' G(\mathbf{x} - \mathbf{x}') V(\mathbf{x}') \psi(\mathbf{x}') \qquad (6.4)$$

where

$$G(\mathbf{x} - \mathbf{x}') = \frac{2}{(2\pi)^d} \int d\mathbf{p} \; \frac{e^{i\mathbf{p}\cdot(\mathbf{x}'-\mathbf{x})}}{p_0^2 + p^2} \qquad (6.5)$$

If we expand the plane wave in (6.3) in terms of a Sturmian basis set, using equation (4.60), and if we cancel the common factor $p_0^2 + p^2$ from both sides, we obtain still another integral representation of the Schrödinger equation, as well as an alternative representation of the kinetic Green's function. Substituting our Sturmian expansion into (6.3), and canceling the common factor $p_0^2 + p^2$ from both sides, we obtain an integral form of the many-particle Schrödinger equation, from which the kinetic energy term has vanished because of the special properties of Sturmian basis sets:

$$\psi^t(\mathbf{p}) = -\frac{1}{p_0^2} \sum_\nu \phi_\nu^t(\mathbf{p}) \int d\mathbf{x} \; \phi_\nu^*(\mathbf{x}) V(\mathbf{x}) \psi(\mathbf{x}) \qquad (6.6)$$

or (taking the Fourier transform of (6.6)),

$$\psi(\mathbf{x}) = -\frac{1}{p_0^2} \sum_\nu \phi_\nu(\mathbf{x}) \int d\mathbf{x}' \; \phi_\nu^*(\mathbf{x}') V(\mathbf{x}') \psi(\mathbf{x}') \qquad (6.7)$$

If we now compare equation (6.7) with (6.5) we can make the identification:

$$G(\mathbf{x} - \mathbf{x}') = \frac{1}{p_0^2} \sum_\nu \phi_\nu(\mathbf{x}) \phi_\nu^*(\mathbf{x}') \qquad (6.8)$$

In this expansion of the kinetic Green's function, the set of functions $\phi_\nu(\mathbf{x})$ can be any Sturmian basis set whatever, i.e. any set of functions satisfying

$$\left[\Delta - p_0^2 \right] \phi_\nu(\mathbf{x}) = 2\beta_\nu V_0(\mathbf{x}) \phi_\nu(\mathbf{x}) \qquad (6.9)$$

where β_ν is chosen in such a way as to give all the members of the set the same energy, regardless of their quantum numbers. In addition, the members of the set must be normalized according to equation (4.54), so that the weighted orthonormality relations will be fulfilled in reciprocal space. Apart from these conditions defining the set of Sturmian basis functions there are no other restrictions; and V_0 may be chosen in any way that is convenient. Thus the expansion of the kinetic Green's function shown in equation (6.8) is very flexible and general.

If we let $\psi^{(i)}$ stand for the ith-order iterated solution of equation (6.7), then the successive orders are linked by the set of equations:

$$\psi^{(1)}(\mathbf{x}) = -\frac{1}{p_0^2} \sum_\nu \phi_\nu(\mathbf{x}) \int dx'\, \phi_\nu^*(\mathbf{x}') V(\mathbf{x}') \psi^{(0)}(\mathbf{x}')$$

$$\psi^{(2)}(\mathbf{x}) = -\frac{1}{p_0^2} \sum_\nu \phi_\nu(\mathbf{x}) \int dx'\, \phi_\nu^*(\mathbf{x}') V(\mathbf{x}') \psi^{(1)}(\mathbf{x}')$$

$$\vdots \quad \vdots \quad \vdots \tag{6.10}$$

from which it follows that

$$\psi^{(2)}(\mathbf{x}) = -\frac{1}{p_0^3} \sum_{\nu',\nu} \phi_{\nu'}(\mathbf{x}) T_{\nu',\nu} \int dx'\, \phi_\nu^*(\mathbf{x}') V(\mathbf{x}') \psi^{(0)}(\mathbf{x}') \tag{6.11}$$

where

$$T_{\nu',\nu} \equiv -\frac{1}{p_0} \int dx\, \phi_{\nu'}^*(\mathbf{x}) V(\mathbf{x}) \phi_\nu(\mathbf{x}) \tag{6.12}$$

If we let

$$\psi^{(i)}(\mathbf{x}) = \sum_\nu \phi_\nu(\mathbf{x}) B_\nu^{(i)} \tag{6.13}$$

then equations (6.10) and (6.11) imply that

$$B_{\nu'}^{(i)} = \frac{1}{p_0} \sum_\nu T_{\nu',\nu} B_\nu^{(i-1)} \tag{6.14}$$

If

$$\psi(\mathbf{x}) = \sum_\nu \phi_\nu(\mathbf{x}) B_\nu \tag{6.15}$$

is an exact solution of (6.7), then the set of coefficients will be the same in all orders and not only the usual secular equation

$$\sum_\nu [T_{\nu',\nu} - p_0 \delta_{\nu',\nu}] B_\nu = 0 \tag{6.16}$$

will hold, but also the higher order equations:

$$\sum_{\nu} \left[T^i_{\nu',\nu} - p^i_0 \delta_{\nu',\nu} \right] B_\nu = 0 \tag{6.17}$$

where T^i denotes the matrix T raised to the ith power. If the function $\psi(\mathbf{x})$ in (6.15) is not an exact solution (because of truncation of the basis set), then the requirements on B_ν imposed by the secular equations (6.17) will differ, depending on the order.

Iteration of the many-center one-electron wave equation; Sum rules

In Chapter 5 we saw that solutions to the one-electron Schrödinger equation for an electron moving in a many-center Coulomb potential of the form

$$v(\mathbf{x}) = -\sum_a \frac{Z_a}{|\mathbf{x} - \mathbf{X}_a|} \tag{6.18}$$

can be built up in momentum space from basis functions of the form:

$$\xi^t_\tau(\mathbf{p}) \equiv \sqrt{\frac{Z_a}{k_\mu n}} \; e^{-i\mathbf{p}\cdot\mathbf{X}_a} \chi^t_{n,l,m}(\mathbf{p}) \tag{6.19}$$

where we have dropped the electron index j, and

$$\chi^t_{n,l,m}(\mathbf{p}) = M(p) Y_{n-1,l,m}(\mathbf{u}) \tag{6.20}$$

with

$$M(p) = \frac{4k_\mu^{5/2}}{(k_\mu^2 + p^2)^2} \tag{6.21}$$

The momentum-space wave functions of an electron in the many-center potential satisfy the integral equation:

$$\varphi^t_\mu(\mathbf{p}) = \sum_\tau \xi^t_\tau(\mathbf{p}) \int d^3p' \; \left(\frac{k_\mu^2 + p'^2}{2k_\mu^2} \right) \xi^{t*}_\tau(\mathbf{p}') \varphi^t_\mu(\mathbf{p}') \tag{6.22}$$

which will have a solution of the form

$$\varphi^t_\mu(\mathbf{p}) = \sum_\tau \xi^t_\tau(\mathbf{p}) C_{\tau,\mu} \tag{6.23}$$

provided that k_μ and $C_{\tau,\mu}$ are the roots and eigenvectors of the secular equation:

$$\sum_\tau \left[K_{\tau',\tau} - k_\mu \delta_{\tau',\tau} \right] C_{\tau,\mu} = 0 \qquad (6.24)$$

where

$$K_{\tau',\tau} \equiv \int d^3p \left(\frac{k_\mu^2 + p^2}{2k_\mu} \right) \xi_{\tau'}^{t*}(\mathbf{p}) \xi_\tau^t(\mathbf{p}) \qquad (6.25)$$

We also saw that the matrix $K_{\tau',\tau}$ can be rewritten in the form

$$K_{\tau',\tau} = \sqrt{\frac{Z_{a'} Z_a}{n'n}} \int d\Omega \; e^{i\mathbf{p}\cdot(\mathbf{X}_{a'} - \mathbf{X}_a)} Y^*_{n'-1,l',m'}(\mathbf{u}) Y_{n-1,l,m}(\mathbf{u}) \qquad (6.26)$$

Just as iteration of (6.7) leads to the higher-order secular equations (6.17), analogously the iteration of (6.22) leads to the higher-order secular equations

$$\sum_\tau \left[K_{\tau',\tau}^i - k_\mu^i \delta_{\tau',\tau} \right] C_{\tau,\mu} = 0 \qquad (6.27)$$

If the basis set is truncated, the requirements imposed on the coefficients $C_{\tau,\mu}$ and the eigenvalues k_μ will depend on the order i. Koga and Matsuhashi [190-192] have shown that the second-iterated many-center secular equation is capable of giving more accurate results than can be obtained with the first-order secular equation. This higher accuracy resulted from the fact that in evaluating the matrix K^2, Koga and Matsuhashi made use of sum rules which implicitly involved basis functions which were not present in their truncated basis set. The impressively accurate results of of a calculation on H_2^+ by these authors are shown in Table 6.1. When $i = 2$, (6.27) becomes:

$$\sum_\tau \left[K_{\tau',\tau}^2 - k_\mu^2 \delta_{\tau',\tau} \right] C_{\tau,\mu} = 0 \qquad (6.28)$$

Equation (6.28) can also be expressed in the form:

$$\sum_\tau K_{\tau',\tau}^2 C_{\tau,\mu} = k_\mu^2 C_{\tau',\mu} \qquad (6.29)$$

and analogously the $i = 1$ secular equation

$$\sum_\tau \left[K_{\tau',\tau} - k_\mu \delta_{\tau',\tau} \right] C_{\tau,\mu} = 0 \qquad (6.30)$$

can be expressed in the form:

$$\sum_{\tau} K_{\tau',\tau} C_{\tau,\mu} = k_\mu C_{\tau',\mu} \tag{6.31}$$

Combining (6.29) and (6.31), we obtain yet another secular equation:

$$\sum_{\tau} \left[K^2_{\tau',\tau} - k_\mu K_{\tau',\tau} \right] C_{\tau,\mu} = 0 \tag{6.32}$$

All these alternative secular equations contain the same information when the basis set is complete; but when the basis set is truncated, they differ from each other; and we can ask which of the possible secular equations then gives the best approximate solutions. It turns out that equation (6.32) yields the most accurate solutions when the basis set is truncated. Some understanding of this fact may be obtained by considering the many-center one-electron problem in direct space.

The many-center problem in direct space

We would like to solve the one-electron wave equation

$$\left[-\frac{1}{2}\Delta + \frac{1}{2}k^2_\mu + v(\mathbf{x}) \right] \varphi_\mu(\mathbf{x}) = 0 \tag{6.33}$$

where $v(\mathbf{x})$ is the many-center nuclear attraction potential shown in equation (6.18). To do this, we can expand the wave function in terms of hydrogenlike Sturmian basis functions located on the various centers:

$$\varphi_\mu(\mathbf{x}) = \sum_{\tau} \xi_\tau(\mathbf{x}) C_{\tau,\mu} \tag{6.34}$$

where τ stands for the set of indices $\{a, n, l, m\}$ and where

$$\xi_\tau(\mathbf{x}) \equiv \sqrt{\frac{Z_a}{k_\mu n}} \, \chi_{n,l,m}(\mathbf{x} - \mathbf{X}_a) \tag{6.35}$$

The momentum-space basis functions shown in equation (6.19) are the Fourier transforms of the direct-space basis functions shown in (6.35).

Substituting the expansion (6.34) into the wave equation (6.33) we obtain

$$\sum_\tau \left[-\frac{1}{2}\Delta + \frac{1}{2}k_\mu^2 + v(\mathbf{x}) \right] \xi_\tau(\mathbf{x})C_{\tau,\mu} = 0 \qquad (6.36)$$

Next we multiply from the left by a complex conjugate function from the basis set and integrate over the electron's coordinates:

$$\sum_\tau \int d^3x\, \xi_{\tau'}^*(\mathbf{x}) \left[-\frac{1}{2}\Delta + \frac{1}{2}k_\mu^2 + v(\mathbf{x}) \right] \xi_\tau(\mathbf{x})C_{\tau,\mu} = 0 \qquad (6.37)$$

Since

$$\xi_\tau(\mathbf{x}) = \frac{1}{(2\pi)^{3/2}} \int d^3p\, e^{i\mathbf{p}\cdot\mathbf{x}}\xi_\tau^t(\mathbf{p}) \qquad (6.38)$$

it follows that

$$\left(-\frac{1}{2}\Delta + \frac{1}{2}k_\mu^2 \right) \xi_\tau(\mathbf{x}) = \frac{1}{(2\pi)^{3/2}} \int d^3p\, e^{i\mathbf{p}\cdot\mathbf{x}} \left(\frac{k_\mu^2 + p^2}{2} \right) \xi_\tau^t(\mathbf{p}) \qquad (6.39)$$

Taking the Fourier transform of (6.39), we obtain:

$$\left[\left(-\frac{1}{2}\Delta + \frac{1}{2}k_\mu^2 \right) \xi_\tau \right]^t (\mathbf{p})$$

$$= \int d^3p' \left(\frac{1}{(2\pi)^3} \int d^3x\, e^{i(\mathbf{p}-\mathbf{p}')\cdot\mathbf{x}} \right) \left(\frac{k_\mu^2 + p'^2}{2} \right) \xi_\tau^t(\mathbf{p}')$$

$$= \left(\frac{k_\mu^2 + p^2}{2} \right) \xi_\tau^t(\mathbf{p}) \qquad (6.40)$$

Finally, making use of (4.46) and (6.25), we derive the result:

$$\int d^3x\, \xi_{\tau'}^*(\mathbf{x}) \left[-\frac{1}{2}\Delta + \frac{1}{2}k_\mu^2 \right] \xi_\tau(\mathbf{x})$$

$$= \int d^3p \left(\frac{k_\mu^2 + p^2}{2} \right) \xi_{\tau'}^{t*}(\mathbf{p})\xi_\tau^t(\mathbf{p}) = k_\mu K_{\tau',\tau} \qquad (6.41)$$

Let us next consider the remaining matrix element in equation (6.37):

$$-\int d^3x\, \xi_{\tau'}^*(\mathbf{x})v(\mathbf{x})\xi_\tau(\mathbf{x})$$

$$= \frac{1}{k_\mu}\sqrt{\frac{Z_{a'}Z_a}{n'n}} \sum_{a''} \int d^3x\, \chi_{n',l',m'}^*(\mathbf{x} - \mathbf{X}_{a'}) \frac{Z_{a''}}{|\mathbf{x} - \mathbf{X}_{a''}|} \chi_{n,l,m}(\mathbf{x} - \mathbf{X}_a)$$

$$\qquad (6.42)$$

Letting α stand for the set of indices $\{n, l, m\}$, and making use of equation (5.72), we have:

$$\chi^*_{n',l',m'}(\mathbf{x} - \mathbf{X}_{a'}) = \sum_{\alpha''} \chi^*_{\alpha''}(\mathbf{x} - \mathbf{X}_{a''}) \left[S^{n'l'm'}_{\alpha''}(\mathbf{X}_{a''} - \mathbf{X}_{a'}) \right]^*$$

$$\chi_{n,l,m}(\mathbf{x} - \mathbf{X}_a) = \sum_{\alpha'} \chi^*_{\alpha'}(\mathbf{x} - \mathbf{X}_{a''}) S^{nlm}_{\alpha'}(\mathbf{X}_{a''} - \mathbf{X}_a) \qquad (6.43)$$

Inserting these expansions into (6.42) and making use of the potential-weighted orthonormality relation (4.53), we have

$$- \int d^3x \; \xi^*_{\tau'}(\mathbf{x}) v(\mathbf{x}) \xi_\tau(\mathbf{x})$$

$$= \sqrt{\frac{Z_{a'} Z_a}{n'n}} \sum_{a''} \sum_{\alpha'} \left[S^{n'l'm'}_{\alpha''}(\mathbf{X}_{a''} - \mathbf{X}_{a'}) \right]^* S^{nlm}_{\alpha'}(\mathbf{X}_{a''} - \mathbf{X}_{a'})$$

$$\times \frac{1}{k_\mu} \int d^3x \; \chi^*_{\alpha''}(\mathbf{x} - \mathbf{X}_{a''}) \frac{Z_{a''}}{|\mathbf{x} - \mathbf{X}_{a''}|} \chi^*_{\alpha'}(\mathbf{X}_a - \mathbf{X}_{a''})$$

$$= \sqrt{\frac{Z_{a'} Z_a}{n'n}} \sum_{\tau''} \left[S^{n'l'm'}_{\alpha''}(\mathbf{X}_{a''} - \mathbf{X}_{a'}) \right]^* \frac{Z_{a''}}{n''} S^{nlm}_{\alpha''}(\mathbf{X}_{a''} - \mathbf{X}_a)$$

$$= \sum_{\tau''} K_{\tau',\tau''} K_{\tau'',\tau} \equiv K^2_{\tau',\tau} \qquad (6.44)$$

From (6.41) and (6.44) we can see that the secular equation (6.37) can be written in the form:

$$\sum_\tau \left[K^2_{\tau',\tau} - k_\mu K_{\tau',\tau} \right] C_{\tau,\mu} = 0 \qquad (6.45)$$

which is identical with (6.32). From (6.44), we can also see that the matrix K^2 can be interpreted in terms of nuclear attraction integrals. Whenever methods for calculating these nuclear attraction integrals are available, it is preferable to use them for evaluating K^2, rather than evaluating K^2 by squaring K with a truncated basis set.

An illustrative example

To illustrate the increased accuracy given by equation (6.32) in the case of a truncated basis set, we can consider the the two-center case, where

$$v(\mathbf{x}) = -\frac{Z_1}{|\mathbf{x} - \mathbf{X}_1|} - \frac{Z_2}{|\mathbf{x} - \mathbf{X}_2|} \qquad (6.46)$$

Then from (6.44), (4.53), and (6.46) and we have:

$$K^2_{1,\alpha';1,\alpha} = \frac{Z_1}{\sqrt{n'n}} \left[\frac{Z_1}{n}\delta_{\alpha',\alpha} + \frac{Z_2}{k_\mu} \int d^3x \chi^*_{\alpha'}(\mathbf{x})\chi_\alpha(\mathbf{x})\frac{1}{|\mathbf{x}-\mathbf{R}|} \right] \quad (6.47)$$

where $\mathbf{R} \equiv \mathbf{X}_2 - \mathbf{X}_1$, and similarly,

$$K^2_{2,\alpha';2,\alpha} = \frac{Z_2}{\sqrt{n'n}} \left[\frac{Z_2}{n}\delta_{\alpha',\alpha} + \frac{Z_1}{k_\mu} \int d^3x \chi^*_{\alpha'}(\mathbf{x})\chi_\alpha(\mathbf{x})\frac{1}{|\mathbf{x}+\mathbf{R}|} \right] \quad (6.48)$$

With the help of (6.44) and (5.74), we can also obtain the terms which are off-diagonal in the indices a' and a:

$$\begin{aligned}
K^2_{1,\alpha';2,\alpha} &= \sqrt{\frac{Z_1 Z_2}{n'n}} \left[\frac{Z_1}{n'} + \frac{Z_2}{n} \right] S^\alpha_{\alpha'}(-\mathbf{R}) \\
K^2_{2,\alpha';1,\alpha} &= \sqrt{\frac{Z_1 Z_2}{n'n}} \left[\frac{Z_1}{n} + \frac{Z_2}{n'} \right] S^\alpha_{\alpha'}(\mathbf{R})
\end{aligned} \quad (6.49)$$

while from (5.14) and (5.15),

$$\begin{aligned}
K_{1,\alpha';1,\alpha} &= \frac{Z_1}{n}\delta_{\alpha',\alpha} \\
K_{2,\alpha';2,\alpha} &= \frac{Z_2}{n}\delta_{\alpha',\alpha} \\
K_{1,\alpha';2,\alpha} &= \sqrt{\frac{Z_1 Z_2}{n'n}} S^\alpha_{\alpha'}(-\mathbf{R}) \\
K_{2,\alpha';1,\alpha} &= \sqrt{\frac{Z_1 Z_2}{n'n}} S^\alpha_{\alpha'}(\mathbf{R})
\end{aligned} \quad (6.50)$$

A curve representing $\epsilon = -k_\mu^2/2$ as a function of $R = s/k_\mu$ can be generated by substituting these values of K^2 and K into equation (6.32). Such a curve, with 3 basis functions on each center, is shown in Figure 6.1, compared with the nearly exact results of Koga and Matsuhashi [190-192]. Comparison shows that equation (6.32) yields a more accurate result with 3 basis functions on each center than could be obtained from (6.30) with 15 basis functions on each center. The numerical values of the ground-state energies for H_2^+ obtained using equation (6.32) with 3 basis functions on each center are shown in column a of Table 6.1, compared with the extremely high-precision results of Koga and Matsuhashi (column b) and the best available direct-space results (column c).

Figure 6.1: This figure shows the ground-state energy of H_2^+ (in Hartrees) as a function of the internuclear separation (in Bohrs), compared with the nearly-exact results of Koga and Matsuhashi [190-192] (dots). The first excited-state energies are shown in the upper curve. The two curves were generated by using equation (6.32) with 3 basis functions on each center.

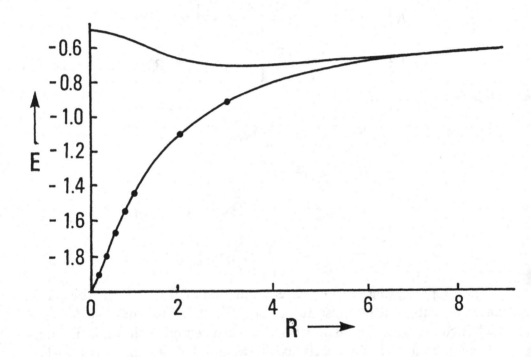

Table 6.1

Nuclear attraction integrals used in equations (6.49) and (6.50). The z-axis is taken to be in the direction of $\mathbf{R}$, and thus $s_3 = s$.

α'	α	$W_{\alpha'}^{\alpha}(\mathbf{R}) \equiv \dfrac{1}{k_\mu} \displaystyle\int d^3x \chi_{\alpha'}^{*}(\mathbf{x})\chi_{\alpha}(\mathbf{x})\dfrac{1}{|\mathbf{x} - \mathbf{R}|}$
$1s$	$1s$	$\dfrac{1}{s} - \dfrac{e^{-2s}}{s}(1 + s)$
$1s$	$2s$	$-\dfrac{1}{2s} + \dfrac{e^{-2s}}{2s}(1 + 2s + 2s^2)$
$1s$	$2p_z$	$\dfrac{1}{s^2} - \dfrac{e^{-2s}}{s^2}(1 + 2s + 2s^2 + s^3)$
$2s$	$2s$	$\dfrac{1}{s} - \dfrac{e^{-2s}}{2s}(2 + 3s + 2s^2 + 2s^3)$
$2s$	$2p_z$	$-\dfrac{3}{2s^2} + \dfrac{e^{-2s}}{2s^2}(3 + 6s + 6s^2 + 4s^3 + 2s^4)$
$2p_z$	$2p_z$	$\dfrac{1}{s} + \dfrac{3}{s^3} - \dfrac{e^{-2s}}{2s^3}(6 + 12s + 14s^2 + 11s^3 + 6s^4 + 2s^5)$

Table 6.2

Ground-state electronic energies of H_2^+ in Hartrees. The results obtained using equation (6.32) with 3 basis functions on each center are shown in column a. The nearly-exact momentum-space results of Koga and Matsuhashi [190-192] are shown in column b, compared with the best position-space results (column c).

R	a	b	c
0.1	−1.9782	−1.978242014	−1.9782421
0.2	−1.9285	−1.928620275	−1.9286202
0.4	−1.8001	−1.800754051	−1.8007539
0.6	−1.6703	−1.671484711	−1.6714846
0.8	−1.5531	−1.554480093	−1.5544801
1.0	−1.4503	−1.451786313	−1.451786313
2.0	−1.1018	−1.102634214	−1.102634214
3.0	−0.9100	−0.910896197	−0.910896197
4.0	−0.7948	−0.796084884	−0.796084884
5.0	−0.7230	−0.724420295	−0.724420295
6.0	−0.6776	−0.678635715	−0.678635715
8.0	−0.6272	−0.627570389	−0.627570389

Exercises

1. Starting with equations (6.44) and (6.46), derive equations (6.47) and (6.48).

2. Show that

$$\frac{1}{k_\mu} \int d^3x \; \chi_{1s}^*(\mathbf{x})\chi_{1s}(\mathbf{x})\frac{1}{|\mathbf{x}-\mathbf{R}|} = \frac{1}{s} - \frac{e^{-2s}}{s}(1+s)$$

(Table 6.1) where $s = k_\mu R$, and where

$$\chi_{1s}(\mathbf{x}) = \left(\frac{k_\mu^3}{\pi}\right)^{1/2} e^{-k_\mu R}$$

What is the limit of the integral as $s \to 0$?

3. Consider two nuclei with charges Z_1 and Z_2. Write down the matrices $K_{\tau',\tau}$ and $K_{\tau',\tau}^2$ for the case where the basis set consists of a single $1s$ atomic orbital localized on each center. Why is the square of the first matrix not equal to the second?

Chapter 7

MOLECULAR STURMIANS

Construction of many-electron Sturmians for molecules

In Chapter 1, we saw that if

$$V_0(\mathbf{x}) = \sum_{j=1}^{N} v(\mathbf{x}_j) \tag{7.1}$$

is the external potential experienced by a collection of N electrons, then it is possible to construct a set of many-electron Sturmian basis functions [41, 45] satisfying

$$\left[-\frac{1}{2}\Delta + \frac{1}{2}p_0^2 + \beta_\nu V_0(\mathbf{x})\right]\phi_\nu(\mathbf{x}) = 0 \tag{7.2}$$

provided that we are able to solve the one-electron equation:

$$\left[-\frac{1}{2}\Delta_j + \frac{1}{2}k_\mu^2 + b_\mu k_\mu v(\mathbf{x}_j)\right]\varphi_\mu(\mathbf{x}_j) = 0 \tag{7.3}$$

The product

$$\phi_\nu(\mathbf{x}) = \varphi_\mu(\mathbf{x}_1)\varphi_{\mu'}(\mathbf{x}_2)...\varphi_{\mu''}(\mathbf{x}_N) \tag{7.4}$$

will then be a solution of (7.2), provided that the subsidiary conditions

$$k_\mu^2 + k_{\mu'}^2 + \ldots + k_{\mu''}^2 = p_0^2 \tag{7.5}$$

and

$$k_\mu b_\mu = k_{\mu'} b_{\mu'} = \ldots = k_{\mu''} b_{\mu''} = \beta_\nu \tag{7.6}$$

are fulfilled. An antisymmetrized wave function built up of products of the form shown in (7.4) will also be a solution of (7.2). The formalism discussed in Chapters 5 and 6 provides us with a method for solving (7.3) in the case where

$$v(\mathbf{x}_j) = -\sum_a \frac{Z_a}{|\mathbf{x}_j - \mathbf{X}_a|} \tag{7.7}$$

i.e. the case where $V_0(\mathbf{x})$ represents the many-center Coulomb attraction potential produced by a set of nuclei with charges Z_a located at the positions $\mathbf{X}_a$. (Notice that we have reintroduced the index j to label the electrons of an N-electron system. We neglected to write this index in Chapters 5 and 6, where we concentrated on the one-electron many-center problem. We have also reintroduced the weighting factor $b_\mu k_\mu$ into the one-electron wave equation, since the one-electron orbitals will now be used to build up a many-electron generalized Sturmian basis set.) We let

$$\varphi_\mu(\mathbf{x}_j) = \sum_\tau \xi_\tau(\mathbf{x}_j) C_{\tau,\mu} \tag{7.8}$$

where

$$\xi_\tau(\mathbf{x}_j) \equiv \sqrt{\frac{Z_a}{k_\mu n}} \, \chi_{nlm}(\mathbf{x}_j - \mathbf{X}_a) \tag{7.9}$$

and where τ represents the set of indices $\{a, n, l, m\}$. Substituting (7.8) into (7.3), we obtain:

$$\sum_\tau \left[-\frac{1}{2}\Delta_j + \frac{1}{2}k_\mu^2 + b_\mu k_\mu v(\mathbf{x}_j) \right] \xi_\tau(\mathbf{x}_j) C_{\tau,\mu} = 0 \tag{7.10}$$

Multiplying (7.10) by $\xi_{\tau'}^*(\mathbf{x}_j)/(k_\mu b_\mu)$ and integrating we obtain

$$\sum_\tau \int d^3 x_j \xi_{\tau'}^*(\mathbf{x}_j) \left[v(\mathbf{x}_j) + \frac{1}{k_\mu b_\mu} \left(-\frac{1}{2}\Delta_j + \frac{1}{2}k_\mu^2 \right) \right] \xi_\tau(\mathbf{x}_j) C_{\tau,\mu} = 0 \tag{7.11}$$

Then from equations (6.41) and (6.44) we obtain the secular equation:

$$\sum_{\tau} \left[K_{\tau',\tau}^2 - b_\mu^{-1} K_{\tau',\tau} \right] C_{\tau,\mu} = 0 \tag{7.12}$$

where

$$K_{\tau',\tau}^2 = -\int d^3x_j \xi_{\tau'}^*(\mathbf{x}_j) v(\mathbf{x}_j) \xi_\tau(\mathbf{x}_j) \tag{7.13}$$

and

$$
\begin{aligned}
K_{\tau',\tau} &= \int d^3x_j \xi_{\tau'}^*(\mathbf{x}_j) \left(-\frac{1}{2}\Delta_j + \frac{1}{2}k_\mu^2 \right) \xi_\tau(\mathbf{x}_j) \\
&= \int d^3p_j \left(\frac{p_j^2 + k_\mu^2}{2k_\mu} \right) \xi_{\tau'}^{*t}(\mathbf{p}_j) \xi_\tau^t(\mathbf{p}_j) \\
&= \sqrt{\frac{Z_{a'} Z_a}{n'n}} S_{n'l'm'}^{nlm}(\mathbf{X}_{a'} - \mathbf{X}_a)
\end{aligned}
\tag{7.14}
$$

Equation (7.12) is identical with equation (6.32), except that k_μ has been replaced by b_μ^{-1}. Having constructed a set of many-electron molecular Sturmian basis functions $\phi_\nu(\mathbf{x})$ with the parameters β_ν chosen in such a way that all functions in the set correspond to the same value of p_0^2, we can normalize them according to the requirement

$$\int dx\, \phi_{\nu'}^*(\mathbf{x}) V_0(\mathbf{x}) \phi_\nu(\mathbf{x}) = -\frac{p_0^2}{\beta_\nu} \delta_{\nu',\nu} \tag{7.15}$$

Alternatively, the many-electron Sturmian basis set may be normalized in momentum space by requiring that

$$\int dp \left(\frac{p^2 + p_0^2}{2p_0^2} \right) \phi_{\nu'}^{*t}(\mathbf{p}) \phi_\nu(\mathbf{p}) = \delta_{\nu',\nu} \tag{7.16}$$

If the functions $\phi_\nu(\mathbf{x})$ are exact solutions to equation (7.2), then the requirements (7.15) and (7.16) are identical; but if the solutions are only approximate, the direct-space and momentum-space normalization requirements are only approximately identical. Having constructed our basis set, we can use it to solve the many-electron Schrödinger equation,

$$\left[-\frac{1}{2}\Delta + \frac{1}{2}p_0^2 + V(\mathbf{x}) \right] \psi(\mathbf{x}) = 0 \tag{7.17}$$

where

$$V(\mathbf{x}) = V_0(\mathbf{x}) + V'(\mathbf{x}) \tag{7.18}$$

In equation (7.18), $V_0(\mathbf{x})$ is the many-center nuclear attraction potential defined by equations (7.1) and (7.7), while

$$V'(\mathbf{x}) = \sum_{j'>j}^{N} \sum_{j=1}^{N} \frac{1}{|\mathbf{x}_j - \mathbf{x}_{j'}|} \tag{7.19}$$

is the interelectron repulsion potential. Letting

$$\psi(\mathbf{x}) = \sum_{\nu} \phi_{\nu}(\mathbf{x}) B_{\nu} \tag{7.20}$$

we obtain the secular equation:

$$\sum_{\nu} [T_{\nu',\nu} - p_0 \delta_{\nu',\nu}] B_{\nu} = 0 \tag{7.21}$$

where

$$\begin{aligned}
T_{\nu',\nu} &\equiv -\frac{1}{p_0} \int d\mathbf{x} \; \phi_{\nu'}^*(\mathbf{x}) V(\mathbf{x}) \phi_{\nu}(\mathbf{x}) \\
&= \delta_{\nu',\nu} \frac{p_0}{\beta_\nu} - \frac{1}{p_0} \int d\mathbf{x} \; \phi_{\nu'}^*(\mathbf{x}) V'(\mathbf{x}) \phi_{\nu}(\mathbf{x})
\end{aligned} \tag{7.22}$$

The factor p_0/β_ν which appears in equation (7.22) can be rewritten by means of the subsidiary conditions, (7.5) and (7.6):

$$\frac{p_0}{\beta_\nu} = \frac{1}{\beta_\nu} \left(k_\mu^2 + k_{\mu'}^2 + \dots \right)^{1/2} = \left(b_\mu^{-2} + b_{\mu'}^{-2} + \dots \right)^{1/2} \tag{7.23}$$

Thus the secular equation, (7.21), becomes

$$\sum_{\nu} \left[T_{\nu',\nu}' + \left(b_\mu^{-2} + b_{\mu'}^{-2} + \dots \right)^{1/2} \delta_{\nu',\nu} - p_0 \delta_{\nu',\nu} \right] B_{\nu} = 0 \tag{7.24}$$

The term representing the nuclear attraction potential is already diagonal in (7.24), and only the interelectron repulsion term

$$T_{\nu',\nu}' = -\frac{1}{p_0} \int d\mathbf{x} \; \phi_{\nu'}^*(\mathbf{x}) V'(\mathbf{x}) \phi_{\nu}(\mathbf{x}) \tag{7.25}$$

remains to be diagonalized.

An illustrative example

As a simple example illustrating the method discussed above, we can consider the molecule H_2 in the approximation where our 2-electron Sturmian basis set consists of the determinential wave functions

$$\phi_1(\mathbf{x}) = \mathcal{N}_1|\varphi_g\varphi_{\bar{g}}| \equiv \varphi_g(\mathbf{x}_1)\varphi_g(\mathbf{x}_2)\frac{\mathcal{N}_1}{\sqrt{2}}\left[\alpha(1)\beta(2) - \alpha(2)\beta(1)\right]$$

$$\phi_2(\mathbf{x}) = \mathcal{N}_2|\varphi_u\varphi_{\bar{u}}| \equiv \varphi_u(\mathbf{x}_1)\varphi_u(\mathbf{x}_2)\frac{\mathcal{N}_2}{\sqrt{2}}\left[\alpha(1)\beta(2) - \alpha(2)\beta(1)\right]$$

$$(7.26)$$

where

$$\begin{aligned}
\varphi_\mu(j) &\equiv \varphi_\mu(\mathbf{x}_j)\alpha(j) \\
\varphi_{\bar{\mu}}(j) &\equiv \varphi_\mu(\mathbf{x}_j)\beta(j)
\end{aligned} \qquad (7.27)$$

and where $\varphi_g(\mathbf{x}_j)$ is the $\sigma g1s$ solution to (7.12) while $\varphi_u(\mathbf{x}_j)$ is the $\sigma^* u1s$ solution. We begin by constructing the matrices K^2 and K using the methods discussed in Chapter 6. As we saw, the matrix elements are functions of the parameter $s = k_\mu R$, where R is the internuclear separation.Choosing a value of s, we then solve equation (7.12) and obtain b_g and b_u for the *gerade* and *ungerade* solutions, as well as the coefficients $C_{\tau,g}$ and $C_{\tau,u}$. If we choose to normalize our basis set in momentum space, then equation (7.16) requires that

$$\mathcal{N}_\mu^2 \int dp \left(\frac{p^2 + p_0^2}{2p_0^2}\right) |\varphi_\mu^t(\mathbf{p}_1)|^2 |\varphi_\mu^t(\mathbf{p}_2)|^2 = 1 \qquad (7.28)$$

From the subsidiary condition (7.5) and from the fact that $b_\mu = b_{\bar{\mu}}$ and $k_\mu = k_{\bar{\mu}}$ we have

$$2k_\mu^2 = p_0^2 \qquad (7.29)$$

Using (7.29) and the fact that

$$p_1^2 + p_2^2 = p^2 \qquad (7.30)$$

we can rewrite (7.28) in the form:

$$\mathcal{N}_\mu^2 \int d^3p_1 \left(\frac{p_1^2 + k_\mu^2}{2k_\mu^2}\right) |\varphi_\mu^t(\mathbf{p}_1)|^2 \int d^3p_2 |\varphi_\mu^t(\mathbf{p}_2)|^2 = 1 \qquad (7.31)$$

The first integral in (7.31) is easy to evaluate, since

$$|\varphi_\mu^t(\mathbf{p}_1)|^2 = \sum_{\tau',\tau} C_{\tau',\mu}^* \xi_{\tau'}^{*t}(\mathbf{p}_1)\xi_\tau^t(\mathbf{p}_1)C_{\tau,\mu} \tag{7.32}$$

Combining (7.32) and (7.14), we obtain:

$$\int d^3p_1 \left(\frac{p_1^2 + k_\mu^2}{2k_\mu}\right)|\varphi_\mu^t(\mathbf{p}_1)|^2 = \sum_{\tau',\tau} C_{\tau',\mu}^* K_{\tau',\tau}C_{\tau,\mu} = (C^{-1}KC)_{\mu,\mu} \tag{7.33}$$

where the matrix C is assumed to be unitary. Similarly,

$$\int d^3p_2 |\varphi_\mu^t(\mathbf{p}_2)|^2 = (C^{-1}MC)_{\mu,\mu} \tag{7.34}$$

where

$$
\begin{aligned}
M_{\tau',\tau} &= \int d^3p_j \xi_{\tau'}^{*t}(\mathbf{p}_j)\xi_\tau^t(\mathbf{p}_j) \\
&= \sqrt{\frac{Z_{a'}Z_a}{n'n}} \int d\Omega e^{i\mathbf{p}_j\cdot(\mathbf{X}_{a'}-\mathbf{X}_a)}(1+u_4)Y_{n'-1,l',m'}^*(\mathbf{u})Y_{n-1,l,m}(\mathbf{u})
\end{aligned}
\tag{7.35}
$$

In equation (7.35) we have made use of the relationship:

$$1+u_4 = \left(\frac{p_j^2 + k_\mu^2}{p_j^2 + k_\mu^2}\right) + \left(\frac{k_\mu^2 + p_j^2}{p_j^2 + k_\mu^2}\right) = \frac{2k_\mu^2}{p_j^2 + k_\mu^2} \tag{7.36}$$

Like the Shibuya-Wulfman integrals, the Sturmian overlap integrals $M_{\tau',\tau}$ shown in equation (7.35) can be evaluated by means of equations (5.53) and (5.54), since

$$\frac{1}{2\pi^2} \int d\Omega \; e^{i\mathbf{p}\cdot\mathbf{R}} \; (1+u_4)u_4^k h_l(u_j) = \mathcal{G}_{k,l}(s)h_l(s_j) \tag{7.37}$$

where

$$\mathcal{G}_{k,l}(s) \equiv \mathcal{F}_{k,l}(s) + \mathcal{F}_{k+1,l}(s) \tag{7.38}$$

Thus we are able to obtain the elements of the matrix M as functions of the parameter $s = k_\mu R$; and the normalization constants can be found by means of the relation

$$\mathcal{N}_\mu^2 \frac{1}{k_\mu}(C^{-1}KC)_{\mu,\mu}(C^{-1}MC)_{\mu,\mu} = 1 \tag{7.39}$$

Knowing the basis functions, we can solve the Sturmian secular equations

$$
\begin{pmatrix} \sqrt{2}b_g^{-1} + T'_{1,1} - p_0 & T'_{1,2} \\ T'_{2,1} & \sqrt{2}b_u^{-1} + T'_{2,2} - p_0 \end{pmatrix} \begin{pmatrix} B_1 \\ B_2 \end{pmatrix} = 0 \qquad (7.40)
$$

where $T'_{\nu',\nu}$ is the matrix of interelectron repulsion integrals. Just as in the case of atoms, this matrix turns out to be independent of p_0; and in fact, it depends only on s. Having chosen a value of s, we can find the corresponding value of p_0 by solving (7.40). The values of energy and internuclear separation can be found from the relationships $E = -p_0^2/2$ and $R = s/k_\mu = \sqrt{2}s/p_0$. By repeating this procedure for many values of s, we can find the energy and wave function of the molecule as functions of R.

Table 7.1: Sturmian overlap integrals
$$\mathcal{G}_{k,l}(s) \equiv \mathcal{F}_{k,l}(s) + \mathcal{F}_{k+1,l}(s)$$

ν	ν'	$\int d\Omega \; e^{i\mathbf{P}\cdot\mathbf{R}}(1+u_4)Y_\nu^*(\mathbf{u})Y_{\nu'}(\mathbf{u})$
$1s$	$1s$	$\mathcal{G}_{0,0}(s)$
$1s$	$2p_j$	$-2i\mathcal{G}_{0,1}(s)s_j \qquad j = 1,2,3$
$1s$	$2s$	$2\mathcal{G}_{1,0}(s)$
$2p_j$	$1s$	$2i\mathcal{G}_{0,1}(s)s_j \qquad j = 1,2,3$
$2p_j$	$2p_j$	$4\mathcal{G}_{0,2}(s)\left[s_j^2 - \dfrac{1}{3}s^2\right] + \dfrac{4}{3}\left[\mathcal{G}_{0,0}(s) - \mathcal{G}_{2,0}(s)\right]$
$2p_j$	$2p_k$	$4\mathcal{G}_{0,2}(s)s_j s_k \qquad j \neq k$
$2p_j$	$2s$	$4i\mathcal{G}_{1,1}(s)s_j$
$2s$	$2s$	$4\mathcal{G}_{2,0}(s)$

Table 7.2

$$\mathcal{G}_{k,l}(s) \equiv \mathcal{F}_{k,l}(s) + \mathcal{F}_{k+1,l}(s)$$

k	l	$\mathcal{G}_{k,l}(s) \qquad s \equiv k_\mu R$
0	0	$\frac{1}{3}(3 + 3s + s^2)e^{-s}$
0	1	$\frac{i}{12}(3 + 3s + s^2)e^{-s}$
0	2	$-\frac{1}{60}(3 + 3s + s^2)e^{-s}$
1	0	$\frac{1}{12}(3 + 3s + 2s^2 + s^3)e^{-s}$
1	1	$\frac{i}{60}(s^2 + s^3)e^{-s}$
1	2	$\frac{1}{360}(3 + 3s - s^3)e^{-s}$
2	0	$\frac{1}{60}(15 + 15s + 5s^2 + s^4)e^{-s}$

Exercises

1. Calculate $\mathcal{G}_{0,0}(s)$, $\mathcal{G}_{1,0}(s)$, and $\mathcal{G}_{0,1}(s)$, using Table 5.2 and the definition

$$\mathcal{G}_{k,l}(s) \equiv \mathcal{F}_{k,l}(s) + \mathcal{F}_{k+1,l}(s)$$

2. Use the results of Exercise 7.1 to calculate the Sturmian overlap integral

$$I = \int d^3x_j \, \chi_{1s}^*(\mathbf{x}_j)\chi_{1s}(\mathbf{x}_j + \mathbf{R})$$

3. Calculate the integral I of Exercise 7.2 using the ellipsoidal coordinates $\xi = (r_a + r_b)/R$, $\eta = (r_a - r_b)/R$ and ϕ where $r_a^2 = \mathbf{x}_j \cdot \mathbf{x}_j$ and $r_b^2 = (\mathbf{x}_j + \mathbf{R}) \cdot (\mathbf{x}_j + \mathbf{R})$ and where ϕ has its usual meaning. In ellipsoidal coordinates, the volume element is given by

$$d^3x = \frac{R^3}{8} \left(\xi^2 - \eta^2 \right) d\xi d\eta d\phi$$

Compare your answer with the results of Exercise 7.2. Could ellipsoidal coordinates be used to calculate Shibuya-Wulfman integrals?

Chapter 8

RELATIVISTIC EFFECTS

Relativistic hydrogenlike Sturmians

The Dirac equation for an electron moving in the Coulomb field of a nucleus with charge b_μ is given (in the clamped nucleus approximation) by [3]

$$\left[D - \epsilon - \frac{e^2 b_\mu}{r} \right] \chi_\mu(\mathbf{x}) = 0 \tag{8.1}$$

where

$$D \equiv -i\hbar c \boldsymbol{\alpha} \cdot \frac{\partial}{\partial \mathbf{x}} + m_0 c^2 \gamma_0 \tag{8.2}$$

and

$$\boldsymbol{\alpha} \equiv \left(\begin{array}{c|c} 0 & \boldsymbol{\sigma} \\ \hline \boldsymbol{\sigma} & 0 \end{array} \right), \qquad \gamma_0 \equiv \left(\begin{array}{c|c} I & 0 \\ \hline 0 & -I \end{array} \right) \tag{8.3}$$

In equation (8.3), $\boldsymbol{\sigma}$ is a vector whose components are the Pauli spin matrices,

$$\boldsymbol{\sigma} \equiv \left\{ \begin{pmatrix} 0 & 1 \\ 1 & 0 \end{pmatrix}, \begin{pmatrix} 0 & -i \\ i & 0 \end{pmatrix}, \begin{pmatrix} 1 & 0 \\ 0 & -1 \end{pmatrix} \right\} \tag{8.4}$$

while I is a 2×2 unit matrix. If we let $\{b_\mu\}$ be a set of parameters especially chosen so that all of the functions in the set correspond to the same value of the relativistic energy ϵ [42], then the set of functions $\chi_\mu(\mathbf{x})$ might be called a *relativistic hydrogenlike Sturmian basis set*.

119

The 4-component solutions of (8.1) can be written in the form:

$$\chi_\mu(\mathbf{x}) = \chi_{njlM}(\mathbf{x}) = \begin{pmatrix} ig_{njl}(r)\Omega_{jlM}(\theta,\varphi) \\ -f_{njl}(r)\Omega_{j\bar{l}M}(\theta,\varphi) \end{pmatrix} \tag{8.5}$$

where $\bar{l} \equiv 2j - l$ and where the functions [3]

$$\Omega_{jlM}(\theta,\varphi) = \begin{pmatrix} Y_{l,M-\frac{1}{2}}(\theta,\varphi)C^{l,1/2}_{M-1/2,1/2;j,M} \\ Y_{l,M+\frac{1}{2}}(\theta,\varphi)C^{l,1/2}_{M+1/2,-1/2;j,M} \end{pmatrix} \tag{8.6}$$

are *spherical spinors*, built up from spherical harmonics and 2-component spinors, combined with appropriate Clebsch-Gordan coefficients. The spherical spinors are simultaneous eigenfunctions of J^2, L^2 and J_z, where

$$\mathbf{J} \equiv \mathbf{L} + \frac{1}{2}\sigma \tag{8.7}$$

and where j, l and M are the quantum numbers labeling the eigenfunctions of the three operators. When $j = l + \frac{1}{2}$,

$$\Omega_{jlM}(\theta,\varphi) = \begin{pmatrix} \sqrt{\dfrac{l+M+\frac{1}{2}}{2l+1}}\ Y_{l,M-\frac{1}{2}}(\theta,\varphi) \\ \sqrt{\dfrac{l-M+\frac{1}{2}}{2l+1}}\ Y_{l,M+\frac{1}{2}}(\theta,\varphi) \end{pmatrix} \tag{8.8}$$

while when $j = l - \frac{1}{2}$,

$$\Omega_{jlM}(\theta,\varphi) = \begin{pmatrix} -\sqrt{\dfrac{l-M+\frac{1}{2}}{2l+1}}\ Y_{l,M-\frac{1}{2}}(\theta,\varphi) \\ \sqrt{\dfrac{l+M+\frac{1}{2}}{2l+1}}\ Y_{l,M+\frac{1}{2}}(\theta,\varphi) \end{pmatrix} \tag{8.9}$$

The radial functions corresponding to the large and small components of $\chi_{njlM}(\mathbf{x})$ are

$$g_{njl}(r) = -\mathcal{N}r^{\gamma-1}e^{-b_\mu r/(Na_0)}$$

$$\times \left\{ n_r F(-n_r+1|2\gamma+1|\frac{2b_\mu r}{Na_0}) - (N-\kappa)F(-n_r|2\gamma+1|\frac{2b_\mu r}{Na_0}) \right\}$$

(8.10)

and

$$f_{njl}(r) = -\mathcal{N}\mathcal{R}r^{\gamma-1}e^{-b_\mu r/(Na_0)}$$

$$\times \left\{ n_r F(-n_r+1|2\gamma+1|\frac{2b_\mu r}{Na_0}) + (N-\kappa)F(-n_r|2\gamma+1|\frac{2b_\mu r}{Na_0}) \right\}$$

(8.11)

where

$$\kappa = \begin{cases} -(j+\frac{1}{2}) & j = l+\frac{1}{2} \\ \\ j+\frac{1}{2} & j = l-\frac{1}{2} \end{cases}$$

(8.12)

$$\gamma \equiv \sqrt{\kappa^2 - \left(\frac{b_\mu e^2}{\hbar c}\right)^2}$$

(8.13)

$$n_r \equiv n - |\kappa|$$

(8.14)

$$N \equiv \sqrt{n^2 - 2n_r(|\kappa| - \gamma)}$$

(8.15)

$$\mathcal{R} \equiv \sqrt{\frac{m_0 c^2 - \epsilon_{nj}}{m_0 c^2 + \epsilon_{nj}}}$$

(8.16)

and where $F(a|b|x)$ is a confluent hypergeometric function:

$$F(a|b|x) \equiv 1 + \frac{ax}{b} + \frac{a(a+1)x^2}{b(b+1)2!} + \cdots$$

(8.17)

The energies of the relativistic hydrogenlike Sturmians are related to the effective nuclear charges b_μ by

$$\epsilon = \frac{m_0 c^2}{\sqrt{1 + \left\{ \frac{b_\mu e^2}{\hbar c(\gamma+n_r)} \right\}^2}}$$

(8.18)

Like the non-relativistic hydrogenlike Sturmians, the relativistic functions shown above obey a potential-weighted orthonormality relation.
We can see this in the following way: Making use of the self-adjointness
of the operator D (equation (8.2)) we can write:

$$[D - \epsilon] \chi_\mu(\mathbf{x}) = b_\mu \frac{e^2}{r} \chi_\mu(\mathbf{x})$$

$$\chi_{\mu'}^\dagger(\mathbf{x}) [D - \epsilon] = b_{\mu'} \chi_{\mu'}^\dagger(\mathbf{x}) \frac{e^2}{r} \tag{8.19}$$

where $\chi_\mu(\mathbf{x})$ and $\chi_{\mu'}(\mathbf{x})$ are two solutions to (8.1). By multiplying by
the appropriate functions, integrating over coordinates, and subtracting
the second equation in (8.20) from the first, we can show that [42]

$$(b_\mu - b_{\mu'}) \int d^3x \; \chi_{\mu'}^\dagger(\mathbf{x}) \frac{e^2}{r} \chi_\mu(\mathbf{x}) = 0 \tag{8.20}$$

and therefore if $b_\mu \neq b_{\mu'}$

$$\int d^3x \; \chi_{\mu'}^\dagger(\mathbf{x}) \frac{e^2}{r} \chi_\mu(\mathbf{x}) = 0 \tag{8.21}$$

If we combine this weighted orthogonality property with respect to the
principal quantum numbers with the orthonormality properties of the
spherical spinors, we can write:

$$\int d^3x \; \chi_{\mu'}^\dagger(\mathbf{x}) \frac{e^2}{r} \chi_\mu(\mathbf{x}) = \delta_{\mu',\mu} \frac{\epsilon}{b_\mu} \tag{8.22}$$

where we have normalized our basis set in such a way that

$$\int d^3x \; \chi_\mu^\dagger(\mathbf{x}) \frac{e^2}{r} \chi_\mu(\mathbf{x}) = \frac{\epsilon}{b_\mu} \tag{8.23}$$

The relativistic hydrogenlike Sturmian basis functions can be used to
build up solutions to the Dirac equation in a non-Coulomb potential.
If we wish to solve the equation

$$[D - \epsilon + \lambda V(\mathbf{x})] \psi(\mathbf{x}) = 0 \tag{8.24}$$

we can represent the wave function as a linear combination of isoenergetic solutions to equation (8.1):

$$\psi(\mathbf{x}) = \sum_{\mu} \chi_{\mu}(\mathbf{x}) B_{\mu} \qquad (8.25)$$

Substituting (8.25) into (8.24), we obtain:

$$\sum_{\mu} [D - \epsilon + \lambda V(\mathbf{x})] \chi_{\mu}(\mathbf{x}) B_{\mu} = 0 \qquad (8.26)$$

If we now make use of (8.1), we can rewrite (8.26) in the form:

$$\sum_{\mu} \left[b_{\mu} \frac{e^2}{r} + \lambda V(\mathbf{x}) \right] \chi_{\mu}(\mathbf{x}) B_{\mu} = 0 \qquad (8.27)$$

The next step is to multiply (8.27) from the left by an adjoint function from our relativistic Sturmian basis set and to integrate over the coordinates, making use of the potential-weighted orthonormality relation, (8.22). This gives us a secular equation of the form:

$$\sum_{\mu} \left[V_{\mu',\mu} + \frac{\epsilon}{\lambda} \delta_{\mu',\mu} \right] B_{\mu} = 0 \qquad (8.28)$$

where

$$V_{\mu',\mu} \equiv \int d^3x \; \chi_{\mu'}^{\dagger}(\mathbf{x}) V(\mathbf{x}) \chi_{\mu}(\mathbf{x}) \qquad (8.29)$$

We can solve the Sturmian secular equation in the following way: First we pick a value of ϵ. We then calculate values of b_{μ}, normalize the basis functions, and we calculate $V_{\mu',\mu}$. The next step is to diagonalize the matrix $V_{\mu',\mu}$. This gives us a spectrum of ϵ/λ values, and hence a spectrum of λ values for which the chosen ϵ is the relativistic energy of the system. One of these λ values corresponds to the ground state of the system, another to the first excited state, and so on. We can repeat this procedure for other values of ϵ, and we can construct by interpolation a set of curves showing $\epsilon(\lambda)$ for the ground state, the first excited state, and also for higher states. Such a set of curves is shown in Figure 8.1 for the case where $\lambda = e^2 Z$ and

$$V(\mathbf{x}) = \frac{e^{-r}}{r} \qquad (8.30)$$

Figure 8.1: This figure shows the ground state and the first few excited states with $l = 0$ and $j = 1/2$ for an electron moving in the screened Coulomb potential shown in equation (8.30). The relativistic energies, from which the electron's rest energy has been subtracted, are expressed in Hartrees, and they are shown as functions of Z. The figure is taken from Avery and Antonsen, reference [38].

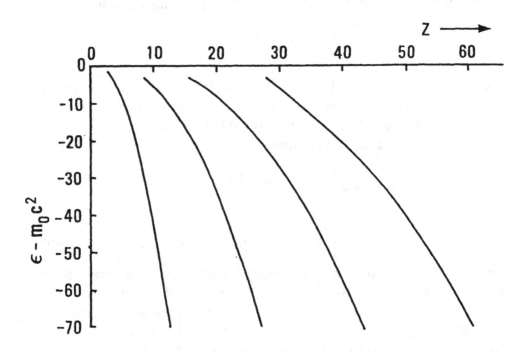

Relativistic many-electron Sturmians

If we wish to construct relativistic many-electron Sturmian basis functions, we must reintroduce an index j to label the electrons in an N-electron system. Thus (8.1) and (8.2) become respectively

$$\left[D_j - \epsilon_\mu - \frac{e^2 \beta_\nu}{r_j} \right] \chi_\mu(\mathbf{x}_j) = 0 \qquad (8.31)$$

(where we have let $b_\mu \to \beta_\nu$) and

$$D_j \equiv -i\hbar c \boldsymbol{\alpha}_j \cdot \frac{\partial}{\partial \mathbf{x}_j} + m_0 c^2 (\gamma_0)_j \qquad (8.32)$$

and where ϵ_μ is given by (8.19). If we let

$$D \equiv \sum_{j=1}^{N} D_j \qquad (8.33)$$

and

$$V_0(\mathbf{x}) = -\sum_{j=1}^{N} \frac{e^2}{r_j} \qquad (8.34)$$

then the function

$$\phi_\nu(\mathbf{x}) = \chi_\mu(\mathbf{x}_1) \chi_{\mu'}(\mathbf{x}_2) ... \chi_{\mu''}(\mathbf{x}_N) \qquad (8.35)$$

will satisfy

$$[D - E + \beta_\nu V_0(\mathbf{x})] \phi_\nu(\mathbf{x}) = 0 \qquad (8.36)$$

provided that

$$\epsilon_\mu + \epsilon_{\mu'} + ... + \epsilon_{\mu''} = E \qquad (8.37)$$

because

$$
\begin{aligned}
D\phi_\nu(\mathbf{x}) &= [D_1 + D_2 + ...] \chi_\mu(\mathbf{x}_1) \chi_{\mu'}(\mathbf{x}_2) ... \\
&= \left[\epsilon_\mu + \frac{e^2 \beta_\nu}{r_1} + \epsilon_{\mu'} + \frac{e^2 \beta_\nu}{r_2} + ... \right] \chi_\mu(\mathbf{x}_1) \chi_{\mu'}(\mathbf{x}_2) ... \\
&= [E - \beta_\nu V_0(\mathbf{x})] \phi_\nu(\mathbf{x}) \qquad (8.38)
\end{aligned}
$$

An antisymmetrized function, $\phi_\nu(\mathbf{x}) = |\chi_\mu \chi_{\mu'} ... \chi_{\mu''}|$, can also be seen to satisfy (8.36). If we let

$$V'(\mathbf{x}) \equiv \sum_{j'>j}^{N} \sum_{j=0}^{N} \frac{e^2}{|\mathbf{x}_j - \mathbf{x}_{j'}|} \tag{8.39}$$

then the Dirac-Coulomb equation for an N-electron atom with nuclear charge Z can be written in the form

$$[D - E + ZV_0(\mathbf{x}) + V'(\mathbf{x})]\,\psi(\mathbf{x}) = 0 \tag{8.40}$$

In order to solve (8.40), we expand the wave function in terms of our N-electron Sturmian basis set:

$$\psi(\mathbf{x}) = \sum_\nu \phi_\nu(\mathbf{x}) B_\nu \tag{8.41}$$

Substituting (8.41) into (8.40), making use of (8.36), and taking the scalar product with an adjoint function in our basis set, we obtain

$$\sum_\nu \int d\mathbf{x}\; \phi_{\nu'}^\dagger(\mathbf{x})\,[(Z - \beta_\nu)V_0(\mathbf{x}) + V'(\mathbf{x})]\,\phi_\nu(\mathbf{x}) B_\nu = 0 \tag{8.42}$$

We now let

$$V'_{\nu',\nu} \equiv \int d\mathbf{x}\; \phi_{\nu'}^\dagger(\mathbf{x}) V'(\mathbf{x}) \phi_\nu(\mathbf{x}) \tag{8.43}$$

and we normalize our basis set in such a way that

$$-\int d\mathbf{x}\; V_0(\mathbf{x}) |\phi_\nu(\mathbf{x})|^2 = 1 \tag{8.44}$$

A potential-weighted orthonormality relation can then be established from the self-adjointness of the operator $[D - E]$ in a manner similar to equations (8.20)-(8.24):

$$-\int d\mathbf{x}\; \phi_{\nu'}^\dagger(\mathbf{x}) V_0(\mathbf{x}) \phi_\nu(\mathbf{x}) = \delta_{\nu',\nu} \tag{8.45}$$

Thus we obtain the secular equation:

$$\sum_\nu \left[V'_{\nu',\nu} + (\beta_\nu - Z)\delta_{\nu',\nu} \right] B_\nu = 0 \tag{8.46}$$

In order to solve these equations, we begin by picking a value of E and a set of configurations $\nu = \{\mu, \mu', ..., \mu''\}$. Then for each configuration in our basis set, we find the value of β_ν for which (8.37) is satisfied. This can be done by constructing a graph of $E(\beta_\nu)$ and interpolating, using (for example) Mathematica. We then find the set of one-electron atomic orbitals which satisfy (8.31), and we construct $V'_{\nu',\nu}$ and J_ν. Diagonalization of $V'_{\nu',\nu}$ then yields a set of Z values for which the secular equation, (8.46), is satisfied; and the smallest of these values corresponds to the ground state. By repeating this procedure for other values of E, we can obtain the energy and wave function of the atom as functions of Z, as illustrated in Table 8.1.

The relativistic many-center one-electron problem

In Chapter 7, we discussed the use of many-center many-electron Sturmian basis sets. It is interesting to ask whether analogous methods can be used for relativistic calculations on molecules. To answer this question, let us begin by considering the many-center one-electron Dirac equation,

$$[D_j - \epsilon_\mu + \beta_\nu v(\mathbf{x}_j)]\, \varphi_\mu(\mathbf{x}_j) = 0 \qquad (8.47)$$

where D_j is defined by (8.32), and

$$v(\mathbf{x}_j) = -\sum_a \frac{Z_a}{|\mathbf{x}_j - \mathbf{X}_a|} \qquad (8.48)$$

We can try to build up the molecular orbital $\varphi_\mu(\mathbf{x}_j)$ from basis functions of the form

$$\xi_\tau(\mathbf{x}_j) = \chi_{njlM}(\mathbf{x}_j - \mathbf{X}_a) \qquad (8.49)$$

where τ stands for the set of indices $\{a, n, j, l, M\}$ and where the functions χ_{njlM} are solutions to

$$\left[D_j - \epsilon_\mu - \frac{e^2 b_{njl}}{r_j}\right] \chi_{njlM}(\mathbf{x}_j) = 0 \qquad (8.50)$$

The constants b_{njl} are chosen in such a way that all the functions χ_{njlM} in the one-electron basis set correspond to the same value of ϵ_μ. From (8.49) and (8.50) it follows that

$$\left[D_j - \epsilon_\mu - \frac{e^2 b_{njl}}{|\mathbf{x}_j - \mathbf{X}_a|} \right] \xi_\tau(\mathbf{x}_j) = 0 \tag{8.51}$$

If we let

$$\varphi_\mu(\mathbf{x}_j) = \sum_\tau \xi_\tau(\mathbf{x}_j) C_{\tau,\mu} \tag{8.52}$$

then, with the help of (8.51), equation (8.47) can be rewritten in the form:

$$\sum_\tau \left[\frac{e^2 b_{njl}}{|\mathbf{x}_j - \mathbf{X}_a|} + \beta_\nu v(\mathbf{x}_j) \right] \xi_\tau(\mathbf{x}_j) C_{\tau,\mu} = 0 \tag{8.53}$$

Multiplying from the left by an adjoint function from the basis set and integrating, we obtain the secular equation:

$$\sum_\tau \int d^3 x_j \xi_{\tau'}^\dagger(\mathbf{x}_j) \left[\frac{e^2 b_{njl}}{|\mathbf{x}_j - \mathbf{X}_a|} + \beta_\nu v(\mathbf{x}_j) \right] \xi_\tau(\mathbf{x}_j) C_{\tau,\mu} = 0 \tag{8.54}$$

If we have a means of evaluating the many-center nuclear attraction integrals which appear in (8.54), then we can proceed as follows: We begin by choosing a value of ϵ_μ. The constants b_{njl} are then determined by inversion of the relationship:

$$\epsilon_\mu = \frac{m_0 c^2}{\sqrt{1 + \left\{ \frac{e^2 b_{njl}}{\hbar c (\gamma + n_r)} \right\}^2}} \tag{8.55}$$

Solution of (8.54) then determines the value of β_ν which corresponds to the chosen value of ϵ_μ for the ground state. Other values of β_ν correspond to the chosen ϵ_μ for the excited states. By repeating this procedure, we are able to obtain ϵ_μ as a function of β_ν, as well as the corresponding molecular orbitals, both for the ground state and for the excited states.

Relativistic many-electron molecular Sturmians

Since we can solve (8.47) by the method just described, it follows that if we let

$$\phi_\nu(\mathbf{x}) = \varphi_\mu(\mathbf{x}_1)\varphi_{\mu'}(\mathbf{x}_2)... \tag{8.56}$$

and

$$E = \epsilon_\mu + \epsilon_{\mu'} + \epsilon_{\mu''} + ... \tag{8.57}$$

then for each configuration $\nu = \{\mu, \mu', \mu'', ...\}$ we can find E as a function of β_ν. The product shown in (8.56) is a solution to

$$[D - E + \beta_\nu V_0(\mathbf{x})]\,\phi_\nu(\mathbf{x}) = 0 \tag{8.58}$$

where D is defined by (8.33) and

$$V_0(\mathbf{x}) = \sum_{j=1}^{N} v(\mathbf{x}_j) \tag{8.59}$$

We can see that it is a solution because, from (8.47) and (8.57)-(8.59) we have:

$$\begin{aligned}
D\phi_\nu(\mathbf{x}) &= [D_1 + D_2 + ...]\,\varphi_\mu(\mathbf{x}_1)\varphi_{\mu'}(\mathbf{x}_2)... \\
&= [\epsilon_\mu - \beta_\nu v(\mathbf{x}_1) + \epsilon_{\mu'} - \beta_\nu v(\mathbf{x}_2) + ...]\,\varphi_\mu(\mathbf{x}_1)\varphi_{\mu'}(\mathbf{x}_2)... \\
&= [E - \beta_\nu V_0(\mathbf{x})]\,\phi_\nu(\mathbf{x})
\end{aligned} \tag{8.60}$$

An antisymmetrized product function will also satisfy (8.58). Thus we are able to construct basis sets which are the relativistic analogs of the many-electron molecular Sturmians discussed in Chapter 7. Choosing a set of functions $\{\phi_\nu(\mathbf{x})\}$ all of whose members correspond to the same value of E, we can use them to build up solutions to the many-center Dirac-Coulomb equation,

$$[D - E + \lambda V_0(\mathbf{x}) + V'(\mathbf{x})]\,\psi(\mathbf{x}) = 0 \tag{8.61}$$

where $V'(\mathbf{x})$ is the interelectron repulsion potential shown in equation (8.39). With the definitions given in equations (8.41)-(8.45), the secular equation becomes:

$$\left[V'_{\nu',\nu} - (\lambda - \beta_\nu)\delta_{\nu',\nu}\right] B_\nu = 0 \tag{8.62}$$

For each value of E, solution of (8.60) gives us values of λ belonging to the ground state and to excited states of the system. By repeating the procedure and by interpolating, we can find the solutions corresponding to $\lambda = 1$, both for the ground state of the system and for the excited states.

Table 8.1: This table shows relativistic and non-relativistic values of the ground-state energies for the 2-electron isoelectronic series of ions for several values of the nuclear charge Z. Values obtained using only one configuration are shown in the column labeled E_{rel1}. A further improvement can be obtained by adding the $|\chi_{2s}\chi_{\bar{2}s}|$ configuration to the basis set, and these two-configuration energies are shown in the column labeled E_{rel2}. The column labeled E_{rel3} shows the results of a much more elaborate calculation by Plante, Johnson and Sapirstein [240]. The non-relativistic values, E_{nr1} and E_{nr2}, correspond respectively to the one-configuration Sturmian approximation, $E_{nr1} = -p_0^2/2$, where $p_0 = \sqrt{2}Z - 0.441942$ and to Clementi's Hartree-Fock values [67].

Z	E_{rel1}	E_{rel2}	E_{rel3}	E_{nr1}	E_{nr2}
10	-93.97	-93.979	-94.028	-93.85	-93.86
20	-388.65	-388.661	-388.712	-387.60	-387.61
30	-892.06	-892.067	-892.116	-881.35	-881.36
40	-1608.88	-1608.895	-1608.932	-1575.10	-1575.11
50	-2556.43	-2556.444	-2556.425	-2468.85	-2468.86

Exercises

1. Show that for $j = 1/2$, $l = 0$, and $M = 1/2$, the 4-component solution to the hydrogenlike Dirac equation can be written in the form:

$$\chi_{n,j,l,M}(\mathbf{x}) = \chi_{n,\frac{1}{2},0,\frac{1}{2}}(\mathbf{x}) = \begin{pmatrix} i g_{n,\frac{1}{2},0}(r) Y_{0,0}(\theta, \phi) \\ 0 \\ \sqrt{\frac{1}{3}} f_{n,\frac{1}{2},0}(r) Y_{1,0}(\theta, \phi) \\ -\sqrt{\frac{2}{3}} f_{n,\frac{1}{2},0}(r) Y_{1,1}(\theta, \phi) \end{pmatrix}$$

   What is the form of the solution corresponding to $j = 1/2$, $l = 0$, and $M = -1/2$?

2. Letting $b_\mu = 1$, find the values of κ, n_r, γ, N, and ϵ_μ for the solutions to (8.1) in the $n = 1$ and $n = 2$ shells.

3. The energies ϵ_μ calculated in Exercise 8.2 include the electron rest energy mc^2 and are expressed in units of mc^2. Subtract the rest energy from the calculated values, and express the results in Hartrees.

Appendix A:
Generalized Slater-Condon rules

A slight complication in the use of many-electron Sturmian basis functions comes from the fact that orthogonality between one-electron orbitals cannot be assumed when one is evaluating matrix elements between configurations corresponding to different values of β_ν. One must then use the generalized Slater-Condon rules. Let us consider two functions ϕ_ν and $\phi_{\nu'}$ in our set of many-electron Sturmian basis functions. These might, for example, be the Slater determinants

$$\phi_{\nu'}(\mathbf{x}) = |P_1 P_2 ... P_N| \qquad (A.1)$$

and

$$\phi_\nu(\mathbf{x}) = |Q_1 Q_2 ... Q_N| \qquad (A.2)$$

If $\beta_\nu \neq \beta_{\nu'}$, then we cannot assume that the two Slater determinants ϕ_ν and $\phi_{\nu'}$ are built up from mutually orthonormal one-electron spin-orbitals; and we must make use of the formalism of generalized Slater-Condon rules in evaluating inter-configurational matrix elements. This formalism was developed by Löwdin, Amos, Hall and others [9,178,216], who showed that it is possible to simplify the evaluation of inter-configurational matrix elements by means of two separate unitary transformations, one mixing the orbitals of the configuration ν, and the other mixing the orbitals of the configuration ν'. When $\beta_\nu \neq \beta_{\nu'}$, the matrix

$$S_{rs} \equiv \int d^3 x_j (P_r^\dagger | Q_s) \qquad (A.3)$$

is in general not diagonal. However, we now make the unitary transformations

$$\hat{P}_i = \sum_r P_r U_{ri} \qquad (A.4)$$

and

$$\hat{Q}_j = \sum_s Q_s W_{sj} \qquad (A.5)$$

We would like to choose the transformation matrices U and W in such a way that

$$\hat{S}_{ij} \equiv \int d^3 x_j (\hat{P}_i^\dagger | \hat{Q}_j) \qquad (A.6)$$

will be diagonal. The authors just mentioned [9,168,206] have shown that we can do this by letting W be the unitary matrix which diagonalizes $S^\dagger S$, i.e. the solution to the secular equation

$$\sum_s \left[(S^\dagger S)_{rs} - \delta_{rs} \lambda_j \right] W_{sj} = 0 \tag{A.7}$$

In matrix notation, equation (A.7) can be rewritten in the form

$$W^\dagger S^\dagger S W = \Lambda \tag{A.8}$$

where Λ is a diagonal matrix whose diagonal elements are λ_j. Since we can also write (A.8) in the form

$$(SW)^\dagger (SW) = \Lambda \tag{A.9}$$

it follows that the diagonal elements λ_j are positive definite. If the λ_j's are also non-zero we can construct the unitary matrix

$$U = SW\Lambda^{-1/2} \tag{A.10}$$

If some of the λ_j's are zero, we can reorder the columns in W so that the vanishing roots occur in a block at the end. The unitary transformation U can then be constructed so that it contains a block mixing the orbitals corresponding to non-zero values of λ_j and satisfying (A.9). The remainder of U can then be constructed in any way desired, provided that the entire matrix is unitary. The unitary transformations defined by (A.7) and (A.10) are just what we need to make $\hat{S}$ diagonal; and we can see that this is the case by the following argument: The adjoint of (A.4) is

$$\hat{P}_i^\dagger = \sum_r U_{ir}^\dagger P_r^\dagger \tag{A.11}$$

If we substitute (A.11) and (A.5) into (A.6), we obtain:

$$\hat{S}_{ij} = \sum_{rs} U_{ir}^\dagger S_{rs} W_{sj} \tag{A.12}$$

or in matrix notation,

$$\hat{S} = U^\dagger S W \tag{A.13}$$

Substituting the adjoint of (A.10) into (A.13), and making use of (A.7), we have

$$\hat{S} = \Lambda^{-1/2} W^\dagger S^\dagger S W = \Lambda^{1/2} \qquad (A.14)$$

which is diagonal because Λ is diagonal. The authors of references [9] and [168] were able to derive generalized Slater-Condon rules in the transformed basis. In our notation, these become:

$$\int dx \phi_{\nu'}^\dagger(\mathbf{x}) \phi_\nu(\mathbf{x}) = |U^\dagger||W| \prod_{j=1}^N \hat{S}_{jj} \qquad (A.15)$$

$$\int dx \phi_{\nu'}^\dagger(\mathbf{x}) \sum_{j=1}^N f(\mathbf{x}_j) \phi_\nu(\mathbf{x})$$

$$= |U^\dagger||W| \sum_{i=1}^N (\prod_{j \neq i}^N \hat{S}_{jj}) \int d^3x_1 (\hat{P}_i^\dagger(1)|f(1)|\hat{Q}_i(1))$$

$$(A.16)$$

$$\int dx \phi_{\nu'}^\dagger(\mathbf{x}) \sum_{i>j}^N \sum_{j=1}^N \frac{1}{r_{ij}} \phi_\nu(\mathbf{x})$$

$$= |U^\dagger||W| \sum_{j>i}^N \sum_{i=1}^N (\prod_{k \neq i,j}^N \hat{S}_{kk})$$

$$\times \int d^3x_1 \int d^3x_2 (\hat{P}_i^\dagger(1)\hat{P}_j^\dagger(2)|\frac{1}{r_{12}}(1 - \mathcal{P}_{12})|\hat{Q}_i(1)\hat{Q}_j(2))$$

$$(A.17)$$

Appendix B:
Coulomb and exchange integrals for atoms

In Chapter 1 (equations (1.36)-(1.48)), we mentioned that in order to construct the interelectron repulsion matrix for atoms, we must be able to evaluate integrals of the form:

$$
\begin{aligned}
J &= \int_0^\infty dr_1 r_1^{2+j_1} e^{-\zeta_1 r_1} \int_0^\infty dr_2 r_2^{2+j_2} e^{-\zeta_2 r_2} \\
&\times \int d\Omega_1 W_1(\hat{\mathbf{x}}_1) \int d\Omega_2 W_2(\hat{\mathbf{x}}_2) \frac{1}{r_{12}} = \sum_l a_l I_l
\end{aligned}
$$

$$(B.1)$$

where

$$
a_l \equiv \int d\Omega_1 W_1(\hat{\mathbf{x}}_1) \int d\Omega_2 W_2(\hat{\mathbf{x}}_2) P_l(\hat{\mathbf{x}}_1 \cdot \hat{\mathbf{x}}_2) \qquad (B.2)
$$

and

$$
I_l \equiv \int_0^\infty dr_1 r_1^{j_1+2} e^{-\zeta_1 r_1} \int_0^\infty dr_2 r_2^{j_2+2} e^{-\zeta_2 r_2} \frac{r_<^l}{r_>^{l+1}} \qquad (B.3)
$$

In equation (B.2), W_1 and W_2 are products of spherical harmonics; and as we shall see below, only a few of the angular integrals a_l are non-zero. In order to evaluate the radial integrals, we can rewrite equation (B.3) in the form:

$$
I_l = \frac{1}{r_1^{l+1}} \int_0^{r_1} dr_2 r_2^{j_2+l+2} e^{-\zeta_2 r_2} + r_1^l \int_{r_1}^\infty dr_2 r_2^{j_2-l+1} e^{-\zeta_2 r_2} \qquad (B.4)
$$

Each of the two terms in (B.4) can be expressed in terms of an incomplete gamma function:

$$
\frac{1}{r_1^{l+1}} \int_0^{r_1} dr_2 r_2^{j_2+l+2} e^{-\zeta_2 r_2} = \frac{1}{(\zeta_2 r_1)^{l+1} \zeta_2^{j_2+2}} \gamma(j_2 + l + 3, \zeta_2 r_1) \qquad (B.5)
$$

and

$$
r_1^l \int_{r_1}^\infty dr_2 r_2^{j_2-l+1} e^{-\zeta_2 r_2} = \frac{(\zeta_2 r_1)^l}{\zeta_2^{j_2+2}} \Gamma(j_2 - l + 2, \zeta_2 r_1) \qquad (B.6)
$$

where

$$
\gamma(a, z) \equiv \int_0^z dt\, t^{a-1} e^{-t} \qquad (B.7)
$$

while

$$\Gamma(a, z) \equiv \int_z^\infty dt \; t^{a-1} e^{-t} \qquad (B.8)$$

The final integration over dr_1 can be performed by means of formulas for the integration of incomplete gamma functions, which can be found in the tables of Gradshteyn and Ryshik [139]. The result is:

$$
\begin{aligned}
&\int_0^\infty dr_1 r_1^{j_1+2} e^{-\zeta_1 r_1} \int_0^\infty dr_2 r_2^{j_2+2} e^{-\zeta_2 r_2} \frac{r_<^l}{r_>^{l+1}} \\
&= \frac{\Gamma(j_1 + j_2 + 5)}{(\zeta_1 + \zeta_2)^{j_1+j_2+5}} \left[\frac{{}_2F_1(1, j_1 + j_2 + 5; j_2 + l + 4; \zeta_2/(\zeta_1 + \zeta_2))}{j_2 + l + 3} \right. \\
&\left. + \frac{{}_2F_1(1, j_1 + j_2 + 5; j_1 + l + 4; \zeta_1/(\zeta_1 + \zeta_2))}{j_1 + l + 3} \right]
\end{aligned}
$$

$$(B.9)$$

Equation (B.9) expresses the radial integral in terms of the hypergeometric function defined by equation (1.43), and thus (B.9) gives the integral in terms of an infinite series. We can, however, transform this result into a polynomial by means of the realtionship [204]:

$$ {}_2F_1(\alpha, \beta; \gamma; z) = (1 - z)^{-\alpha} \; {}_2F_1(\alpha, \gamma - \beta; \gamma; z/(z - 1)) \qquad (B.10)$$

Letting

$$z = \frac{\zeta_2}{\zeta_1 + \zeta_2} \qquad (B.11)$$

so that

$$
\begin{aligned}
\frac{z}{z - 1} &= -\frac{\zeta_2}{\zeta_1} \\
(1 - z)^{-1} &= \frac{\zeta_1 + \zeta_2}{\zeta_1}
\end{aligned}
$$

$$(B.12)$$

we can transform the first hypergeometric function of (B.9):

$$
\begin{aligned}
&{}_2F_1(1, j_1 + j_2 + 5; j_2 + l + 4; \zeta_2/(\zeta_1 + \zeta_2)) \\
&= \frac{\zeta_1 + \zeta_2}{\zeta_1} \; {}_2F_1(1, l - j_1 - 1; j_2 + l + 4; -\zeta_2/\zeta_1)
\end{aligned}
$$

$$(B.13)$$

and similarly,

$$_2F_1(1, j_1 + j_2 + 5; j_1 + l + 4; \zeta_1/(\zeta_1 + \zeta_2))$$

$$= \frac{\zeta_1 + \zeta_2}{\zeta_2} \, _2F_1(1, l - j_2 - 1; j_1 + l + 4; -\zeta_1/\zeta_2)$$

$$(B.14)$$

Thus we obtain the radial integral in terms of a polynomial:

$$\int_0^\infty dr_1 \, r_1^{j_1+2} e^{-\zeta_1 r_1} \int_0^\infty dr_2 \, r_2^{j_2+2} e^{-\zeta_2 r_2} \frac{r_<^l}{r_>^{l+1}}$$

$$= \frac{\Gamma(j_1 + j_2 + 5)}{(\zeta_1 + \zeta_2)^{j_1 + j_2 + 4}} \left[\frac{_2F_1(1, l - j_1 - 1; j_2 + l + 4; -\zeta_2/\zeta_1)}{(j_2 + l + 3)\zeta_1} \right.$$

$$\left. + \frac{_2F_1(1, l - j_2 - 1; j_1 + l + 4; -\zeta_1/\zeta_2)}{(j_1 + l + 3)\zeta_2} \right]$$

$$(B.15)$$

since in all cases of interest, $l - j_1 - 1$ and $l - j_2 - 1$ are either zero or else negative integers. Thus the series defined by equation (1.43) terminates for both terms in (B.15). A particularly simple case occurs when

$$l - j_1 - 1 = 0$$
$$l - j_2 - 1 = 0$$
$$j_1 = j_2 = l - 1$$

$$(B.16)$$

In that case, both hypergeometric functions are equal to 1, and the radial integral reduces to

$$I_l \rightarrow \frac{\Gamma(2l + 3)}{(\zeta_1 + \zeta_2)^{2l+2}} \left[\frac{1}{(2l + 2)\zeta_1} + \frac{1}{(2l + 2)\zeta_2} \right] \qquad (B.17)$$

so that

$$\int_0^\infty dr_1 \, r_1^{l+1} e^{-\zeta_1 r_1} \int_0^\infty dr_2 \, r_2^{l+1} e^{-\zeta_2 r_2} \frac{r_<^l}{r_>^{l+1}} = \frac{(2l + 1)!}{\zeta_1 \zeta_2 (\zeta_1 + \zeta_2)^{2l+1}} \qquad (B.18)$$

The integrals for other values of j_1 and j_2 can be found by differentiating both sides of (B.18) with respect to ζ_1 and ζ_2:

$$\int_0^\infty dr_1\, r_1^{j_1+2} e^{-\zeta_1 r_1} \int_0^\infty dr_2\, r_2^{j_2+2} e^{-\zeta_2 r_2} \frac{r_<^l}{r_>^{l+1}}$$

$$= \left(-\frac{\partial}{\partial\zeta_1}\right)^{j_1-l+1} \left(-\frac{\partial}{\partial\zeta_2}\right)^{j_2-l+1} \frac{(2l+1)!}{\zeta_1\zeta_2(\zeta_1+\zeta_2)^{2l+1}}$$

$$(B.19)$$

and this result can be regarded as an alternative form of (B.15). We mentioned above that only a few of the angular integrals a_l are non-zero. In the case of Coulomb integrals, they take on the form

$$a_l^c \equiv \int d\Omega_1 |Y_{l_1,m_1}(\hat{\mathbf{x}}_1)|^2 \int d\Omega_2 |Y_{l_2,m_2}(\hat{\mathbf{x}}_2)|^2 P_l(\hat{\mathbf{x}}_1 \cdot \hat{\mathbf{x}}_2) \qquad (B.20)$$

while for exchange integrals they have the form:

$$a_l^{ex} \equiv \int d\Omega_1 Y_{l_1,m_1}^*(\hat{\mathbf{x}}_1) Y_{l_2,m_2}(\hat{\mathbf{x}}_1) \int d\Omega_2 Y_{l_2,m_2}^*(\hat{\mathbf{x}}_2) Y_{l_1,m_1}(\hat{\mathbf{x}}_2) P_l(\hat{\mathbf{x}}_1 \cdot \hat{\mathbf{x}}_2)$$

$$(B.21)$$

Since $P_0(\hat{\mathbf{x}}_1 \cdot \hat{\mathbf{x}}_2) = 1$, we can see immediately from (B.20) and (B.21) that

$$a_0^c = 1 \qquad\qquad a_0^{ex} = \delta_{l_1,l_2}\delta_{m_1,m_2} \qquad (B.22)$$

For other values of l, it is helpful to use the relationship,

$$\int d\Omega_2 W_2(\hat{\mathbf{x}}_2) P_l(\hat{\mathbf{x}}_1 \cdot \hat{\mathbf{x}}_2) = \frac{4\pi}{2l+1} O_l\left[W_2(\hat{\mathbf{x}}_1)\right] \qquad (B.23)$$

which follows from equation (3.71). In equation (B.23), the operator O_l is a projection operator corresponding to the angular momentum quantum number l. For the projection to be non-zero, P_l must have the same parity as W_2, from which it follows that

$$a_l^c = 0 \qquad \text{if } l = \text{odd} \qquad (B.24)$$

and that

$$a_l^{ex} = 0 \qquad \text{if } (-1)^l \neq (-1)^{l_1+l_2} \qquad (B.25)$$

Finally, the projection will be zero if l exceeds the order of the homogeneous polynomial which can be formed from W_1 or W_2 by multiplying these angular functions by the appropriate power of r. Therefore

$$a_l^c = 0 \quad \text{if } l > 2l_< \qquad l_< \equiv \begin{cases} l_1 & \text{if } l_1 < l_2 \\ l_2 & \text{if } l_2 < l_1 \end{cases} \qquad (B.26)$$

and

$$a_l^{ex} = 0 \quad \text{if } l > l_1 + l_2 \qquad (B.27)$$

For non-zero cases, the harmonic projection formalism of Chapter 3 may be used to evaluate the angular integrals a_l. Alternatively these integrals may be evaluated by means of Clebsch-Gordan coefficients using the relationships

$$a_l^c = \frac{4\pi}{2l+1} \int d\Omega_1 |Y_{l_1,m_1}(\hat{\mathbf{x}}_1)|^2 Y_{l,0}^*(\hat{\mathbf{x}}_1) \int d\Omega_2 |Y_{l_2,m_2}(\hat{\mathbf{x}}_2)|^2 Y_{l,0}(\hat{\mathbf{x}}_2)$$

$$(B.28)$$

and

$$a_l^{ex} = \frac{4\pi}{2l+1} \int d\Omega_1 Y_{l_1,m_1}^*(\hat{\mathbf{x}}_1) Y_{l_2,m_2}(\hat{\mathbf{x}}_1) Y_{l,m_1-m_2}(\hat{\mathbf{x}}_1)$$

$$\times \int d\Omega_2 Y_{l_2,m_2}^*(\hat{\mathbf{x}}_2) Y_{l_1,m_1}(\hat{\mathbf{x}}_2) Y_{l,m_1-m_2}^*(\hat{\mathbf{x}}_2)$$

$$(B.29)$$

which follow from equation (3.71).

Solutions to the exercises

Exercise 1.1

Use Table 1.2 to show that the one-electron hydrogenlike Sturmian $\chi_{1,0,0,\frac{1}{2}}(\mathbf{x}_j)$ obeys equations (1.11)-(1.13).

Solution

$$\chi_{1,0,0,\frac{1}{2}}(\mathbf{x}_j) = \left(\frac{k_1^3}{\pi}\right)^{1/2} e^{-k_1 r_j} \alpha(j)$$

$$\left[\Delta_j - k_1^2\right] \chi_{1,0,0,\frac{1}{2}}(\mathbf{x}_j) = \left(\frac{k_1^3}{\pi}\right)^{1/2} \alpha(j) \left[k_1^2 - \frac{2k_1}{r_j} - k_1^2\right] e^{-k_1 r}$$

$$= 2\left(\frac{k_1}{Z}\right)\left(-\frac{Z}{r_j}\right) \chi_{1,0,0,\frac{1}{2}}(\mathbf{x}_j)$$

$$\int d\tau_j |\chi_{1,0,0,\frac{1}{2}}(\mathbf{x}_j)|^2 \left(-\frac{Z}{r_j}\right) = -\frac{Zk_1^3}{\pi} \int d^3x_j\, e^{-2k_1 r_j} \frac{1}{r_j}$$

$$= -4Zk_1^3 \int_0^\infty dr_j\, r_j e^{-2k_1 r_j}$$

$$= -\frac{4Zk_1^3}{(2k_1)^2} = -\frac{k_1^2}{(k_1/Z)}$$

$$\int d\tau_j |\chi_{1,0,0,\frac{1}{2}}(\mathbf{x}_j)|^2 = \frac{k_1^3}{\pi} \int d^3x_j\, e^{-2k_1 r_j}$$

$$= 4k_1^3 \int_0^\infty dr_j\, r_j^2 e^{-2k_1 r_j}$$

$$= -\frac{4k_1^3 2!}{(2k_1)^3} = 1$$

Exercise 1.2

Show that if $k_1 = \beta/Z$ and $k_2 = \beta/(2Z)$, then $\chi_{1,0,0,\frac{1}{2}}(\mathbf{x}_j)$ and $\chi_{2,0,0,\frac{1}{2}}(\mathbf{x}_j)$ obey the orthogonality relation

$$\int d\tau_j\, \chi^*_{1,0,0,\frac{1}{2}}(\mathbf{x}_j)\chi_{2,0,0,\frac{1}{2}}(\mathbf{x}_j) = 0$$

Solution

$$\chi_{1,0,0,\frac{1}{2}}(\mathbf{x}_j) = \left(\frac{k_1^3}{\pi}\right)^{1/2} e^{-k_1 r_j} \alpha(j)$$

$$\chi_{2,0,0,\frac{1}{2}}(\mathbf{x}_j) = \left(\frac{k_1^3}{\pi}\right)^{1/2} e^{-k_1 r_j} (1 - k_2 r_j) \alpha(j)$$

$$\int d\tau_j \chi^*_{1,0,0,\frac{1}{2}}(\mathbf{x}_j) \chi_{2,0,0,\frac{1}{2}}(\mathbf{x}_j) = \frac{(k_1 k_2)^{3/2}}{\pi} \int d^3 x_j e^{-(k_1+k_2)r_j} (1 - k_2 r_j)$$

$$= 4(k_1 k_2)^{3/2} \int_0^\infty dr_j \, r_j^2 e^{-(k_1+k_2)r_j} (1 - k_2 r_j)$$

$$= 4(k_1 k_2)^{3/2} \left[\frac{2!}{(k_1 + k_2)^2} - \frac{3! k_2}{(k_1 + k_2)^3}\right]$$

$$= \frac{8(k_1 k_2)^{3/2}}{(k_1 + k_2)^3}(k_1 - 2k_2) = 0$$

From exercise (1.1),

$$\int d\tau_j |\chi^*_{1,0,0,\frac{1}{2}}(\mathbf{x}_j)|^2 = 1$$

while

$$\int d\tau_j |\chi^*_{2,0,0,\frac{1}{2}}(\mathbf{x}_j)|^2 = \frac{k_2^3}{\pi} \int d^3 x_j e^{-2k_2 r_j} (1 - k_2 r_j)^2$$

$$= 4k_2^3 \int_0^\infty dr_j \, r_j^2 e^{-2k_2 r_j} (1 - 2k_2 r_j + k_2^2 r_j^2)$$

$$= 4k_2^3 \left[\frac{2!}{(2k_2)^3} - \frac{2k_2 3!}{(2k_2)^4} + \frac{k_2^2 4!}{(2k_2)^5}\right] = 1$$

Exercise 1.3

Show that if $k_1 = k_2 = k_\mu$, then $\chi_{1,0,0,\frac{1}{2}}(\mathbf{x}_j)$ and $\chi_{2,0,0,\frac{1}{2}}(\mathbf{x}_j)$ obey the potential-weighted orthonormality relation

$$\frac{n}{k_\mu} \int d\tau_j \, \chi^*_{nlms}(\mathbf{x}_j) \frac{1}{r} \chi_{n'l'm's'}(\mathbf{x}_j) = \delta_{n'n}\delta_{l'l}\delta_{m'm}\delta_{s's}$$

Solution

$$\chi_{1,0,0,\frac{1}{2}}(\mathbf{x}_j) = \left(\frac{k_\mu^3}{\pi}\right)^{1/2} e^{-k_\mu r_j}\alpha(j)$$

$$\chi_{2,0,0,\frac{1}{2}}(\mathbf{x}_j) = \left(\frac{k_\mu^3}{\pi}\right)^{1/2} e^{-k_\mu r_j}(1 - k_\mu r_j)\alpha(j)$$

$$
\begin{aligned}
\int d\tau_j \chi_{1,0,0,\frac{1}{2}}^*(\mathbf{x}_j)\frac{1}{r_j}\chi_{2,0,0,\frac{1}{2}}(\mathbf{x}_j) &= \frac{k_\mu^3}{\pi}\int d\tau_j \frac{1}{r_j}e^{-2k_\mu r_j}(1 - k_\mu r_j) \\
&= 4k_\mu^3\int_0^\infty dr_j\, r_j e^{-2k_\mu r_j}(1 - k_\mu r_j) \\
&= 4k_\mu^3\left[\frac{1!}{(2k_\mu)^2} - \frac{2!k_\mu}{(2k_\mu)^3}\right] = 0
\end{aligned}
$$

$$
\begin{aligned}
\frac{1}{k_\mu}\int d\tau_j |\chi_{1,0,0,\frac{1}{2}}(\mathbf{x}_j)|^2\frac{1}{r_j} &= \frac{1}{k_\mu}\left(\frac{k_\mu^3}{\pi}\right)\int d^3x_j e^{-2k_\mu r_j}\frac{1}{r_j} \\
&= 4k_\mu^2\int_0^\infty dr_j\, r_j e^{-2k_\mu r_j} \\
&= 4k_\mu^2\frac{1}{(2k_\mu)^2} = 1
\end{aligned}
$$

$$
\begin{aligned}
\frac{2}{k_\mu}\int d\tau_j |\chi_{2,0,0,\frac{1}{2}}(\mathbf{x}_j)|^2\frac{1}{r_j} &= \frac{2}{k_\mu}\left(\frac{k_\mu^3}{\pi}\right)\int d^3x_j e^{-2k_\mu r_j}(1 - k_\mu r_j)\frac{1}{r_j} \\
&= 8k_\mu^2\int_0^\infty dr_j\, r_j e^{-2k_\mu r_j}(1 - 2k_\mu r_j + k_\mu^2 r_j^2) \\
&= 8k_\mu^2\left[\frac{1!}{(2k_\mu)^2} - \frac{2k_\mu 2!}{(2k_\mu)^3} + \frac{k_\mu^3 3!}{(2k_\mu)^4}\right] = 1
\end{aligned}
$$

Exercise 1.4

Use equations (1.7) and (1.8) to evaluate the direct-space hydrogenlike orbitals χ_{1s}, χ_{2s}, χ_{2p_j}, χ_{3s} and χ_{3p_j}.

Solution

For $l = 0$, $Y_{0,0} = 1/\sqrt{4\pi}$ and (with $t = k_\mu r$)

$$
\begin{aligned}
R_{n,0} &= 2k_\mu^{3/2} e^{-t} F(1 - n|2|2t) \\
&= 2k_\mu^{3/2} e^{-t} \left[1 + \frac{(1-n)}{2} 2t + \frac{(1-n)(2-n)}{2 \cdot 3} \frac{(2t)^2}{2!} + \cdots \right]
\end{aligned}
$$

so that

$$
\chi_{1s} = \left(\frac{k_\mu^3}{\pi} \right)^{3/2} e^{-t}
$$

$$
\chi_{2s} = \left(\frac{k_\mu^3}{\pi} \right)^{3/2} e^{-t} (1 - t)
$$

$$
\chi_{3s} = \left(\frac{k_\mu^3}{\pi} \right)^{3/2} e^{-t} (1 - 2t + \frac{2}{3} t^2)
$$

For $l = 1$,

$$
Y_{1,0} = \sqrt{\frac{3}{4\pi}} \frac{t_3}{t}
$$

and

$$
\begin{aligned}
R_{n,1} &= \frac{2}{3} k_\mu^{3/2} (n^2 - 1)^{1/2} e^{-t} F(2 - n|4|2t) \\
&= \frac{2}{3} k_\mu^{3/2} (n^2 - 1)^{1/2} e^{-t} \left[1 + \frac{(2-n)}{4} 2t + \cdots \right]
\end{aligned}
$$

so that

$$
\chi_{2p_j} = \left(\frac{k_\mu^3}{\pi} \right)^{3/2} e^{-t} t_j
$$

$$
\chi_{3p_j} = \left(\frac{8k_\mu^3}{3\pi} \right)^{3/2} e^{-t} (1 - \frac{t}{2}) t_j
$$

Exercise 2.1

Calculate the integral in equation (2.6) and show that it yields the result shown (2.7).

Solution

$$\frac{1}{p}\int_0^\infty dr\ e^{-k_\mu r}\sin(pr) = \frac{1}{2ip}\int_0^\infty dr\ e^{(-k_\mu+ip)r}\sin(pr) - c.c$$

$$= \frac{1}{2ip}\left[\frac{e^{(-k_\mu+ip)r}}{-k_\mu+ip} - c.c\right]_0^\infty$$

$$= \frac{1}{2ip}\left[\frac{1}{k_\mu-ip} - c.c\right]$$

$$= \frac{1}{p^2+k_\mu^2}$$

Exercise 2.2

Starting with J_{10}, calculate the integrals J_{20} and J_{30} by differentiating with respect to k_μ, as shown in equations (2.8) and (2.9).

Solution

$$J_{1,0} = (p^2+k_\mu^2)^{-1}$$

$$J_{2,0} = -\frac{\partial}{\partial k_\mu}(p^2+k_\mu^2)^{-1} = 2k_\mu(p^2+k_\mu^2)^{-2}$$

$$J_{3,0} = -\frac{\partial}{\partial k_\mu}\left[2k_\mu(p^2+k_\mu^2)^{-2}\right]$$

$$= \left[8k_\mu^2 - 2(p^2+k_\mu^2)\right](p^2+k_\mu^2)^{-3}$$

Exercise 2.3

Starting with J_{10}, use the recursion relations of equation (2.11) to generate J_{21}, J_{31} and J_{32}.

Solution

$$J_{l+1,l} = \frac{2lp}{p^2+k_\mu^2}J_{l,l-1}$$

$$J_{2,1} = \frac{2p}{p^2 + k_\mu^2} J_{1,0} = \frac{2p}{(p^2 + k_\mu^2)^2}$$

$$J_{3,2} = \frac{4p}{p^2 + k_\mu^2} J_{2,1} = \frac{8p^2}{(p^2 + k_\mu^2)^3}$$

$$J_{3,1} = -\frac{\partial}{\partial k_\mu} J_{2,1} = \frac{8pk_\mu}{(p^2 + k_\mu^2)^3}$$

Exercise 2.4

Use the integrals J_{sl} in Table 2.1 to evaluate the Fourier transforms of the direct-space hydrogenlike orbitals χ_{1s}, χ_{2s} and χ_{2p_j}. Show that the transforms correspond to the Solutions of Fock, equation (2.15).

Solution

$$
\begin{aligned}
\chi_{1s}^t(\mathbf{p}) &= \frac{1}{(2\pi)^{3/2}} \int d^3x \, e^{-i\mathbf{p}\cdot\mathbf{x}} \left(\frac{k_\mu^3}{\pi}\right)^{1/2} e^{-k_\mu r} \\
&= \frac{4\pi}{(2\pi)^{3/2}} \left(\frac{k_\mu^3}{\pi}\right)^{1/2} \int_0^\infty dr \, r^2 j_0(pr) e^{-k_\mu r} \\
&= \frac{(2k_\mu^3)^{1/2}}{\pi} J_{2,0} = \left(\frac{4k_\mu^{5/2}}{(p^2 + k_\mu^2)^2}\right) \frac{1}{\sqrt{2\pi}} \\
&= M(p) Y_{0,0,0}(\mathbf{u})
\end{aligned}
$$

$$
\begin{aligned}
\chi_{2s}^t(\mathbf{p}) &= \frac{1}{(2\pi)^{3/2}} \int d^3x \, e^{-i\mathbf{p}\cdot\mathbf{x}} \left(\frac{k_\mu^3}{\pi}\right)^{1/2} e^{-k_\mu r}(1 - k_\mu r) \\
&= \frac{4\pi}{(2\pi)^{3/2}} \left(\frac{k_\mu^3}{\pi}\right)^{1/2} \int_0^\infty dr \, r^2 j_0(pr) e^{-k_\mu r}(1 - k_\mu r) \\
&= \frac{(2k_\mu^3)^{1/2}}{\pi} (J_{2,0} - k_\mu J_{3,0}) \\
&= \left(\frac{4k_\mu^{5/2}}{(p^2 + k_\mu^2)^2}\right) \frac{(-2)}{\sqrt{2\pi}} \left(\frac{k_\mu^2 - p^2}{p^2 + k_\mu^2}\right) \\
&= M(p) Y_{1,0,0}(\mathbf{u})
\end{aligned}
$$

$$\chi^t_{2p_j} = R^t_{2,1}(p)\sqrt{\frac{3}{4\pi}}\frac{p_j}{p}$$

$$R^t_{2,1}(p) = -i\sqrt{\frac{2}{3\pi}}2k^{5/2}_\mu \int_0^\infty dr\ r^3 j_1(pr)e^{-k_\mu r}$$

$$= -i\sqrt{\frac{2}{3\pi}}2k^{5/2}_\mu J_{3,1}$$

$$\chi^t_{2p_j} = -i\sqrt{\frac{2}{3\pi}}2k^{5/2}_\mu \frac{8pk_\mu}{(p^2+k^2_\mu)^3}\sqrt{\frac{3}{4\pi}}\frac{p_j}{p}$$

$$= \left(\frac{4k^{5/2}_\mu}{(p^2+k^2_\mu)^2}\right)\frac{(-2i)}{\sqrt{2\pi}}\left(\frac{2k_\mu p_j}{p^2+k^2_\mu}\right)$$

$$= M(p)Y_{2p_j}(\mathbf{u})$$

Exercise 3.1

Which of the following polynomials in a d-dimensional space are homogeneous? Which are harmonic? (r is the hyperradius.)

1. $x^3_1 + x^3_2$

2. $x^3_1 + x^2_2$

3. $2x^3_1 + x^3_2$

4. $x^3_1 - x^3_2$

5. $x^3_1 - 3x_1 r^2/(d+2)$

6. $x^2_1 x_2 x_3 - r^2 x_2 x_3/(d+4)$

Solution

	homogeneous	harmonic
1	yes	no
2	no	no
3	yes	no
4	yes	yes
5	yes	yes
6	yes	yes

Exercise 3.2

Use equation (3.24) to find expressions analogous to (3.18) for the harmonic decomposition of a 4th-order polynomial, f_4.

Solution

When $n = 4$ and $\nu = 2$, (3.24) becomes

$$
\begin{aligned}
h_0 &= \frac{(d+8-8-2)!!}{4!!(d+8-4-2)!!} \frac{(d+8-8-4)!!}{(d+8-8-4)!!} \Delta^2 f_4 \\
&= \frac{(d-2)!!}{8(d+2)!!} \Delta^2 f_4 \\
&= \frac{1}{8d(d+2)} \Delta^2 f_4
\end{aligned}
$$

When $n = 4$ and $\nu = 1$, (3.24) becomes

$$
\begin{aligned}
h_2 &= \frac{(d+2)!!}{2(d+4)!!} \sum_{j=0}^{1} \frac{(-1)^j (d-2j)!!}{(2j)!!d!!} r^{2j} \Delta^{j+1} f_4 \\
&= \frac{1}{2(d+4)} \left[\Delta f_4 - \frac{r^2}{2d} \Delta^2 f_4 \right]
\end{aligned}
$$

When $n = 4$ and $\nu = 0$, (3.24) becomes

$$
h_4 = \sum_{j=0}^{2} \frac{(-1)^j (d+4-2j)!!}{(2j)!!(d+4)!!} r^{2j} \Delta^j f_4
$$

$$= f_4 - \frac{(d+2)!!r^2}{2(d+4)!!}\Delta f_4 + \frac{d!!r^4}{8(d+4)!!}\Delta^2 f_4$$

$$= f_4 - \frac{r^2}{2(d+4)}\Delta f_4 + \frac{r^4}{8(d+2)(d+4)}\Delta^2 f_4$$

Exercise 3.3

Use equation (3.49) to calculate the normalization factor, in a 4-dimensional space, for the hyperspherical harmonic shown in equation (3.6). Compare this result with Table 2.4.

Solution

$$\mathcal{N}^2 \int d\Omega \left(\frac{x_1^2 - x_2^2}{r^2}\right)^2 = \frac{\mathcal{N}^2}{r^4} \int d\Omega (x_1^4 - 2x_1^2 x_2^2 + x_4^2) = 1$$

For $d = 4$, (3.49) becomes:

$$r^{-n} \int d\Omega \prod_{j=1}^{4} x_j^{n_j} = \frac{4\pi^2}{(n+2)!!} \prod_{j=1}^{4} (n_j - 1)!!$$

so that

$$\frac{1}{r^4} \int d\Omega \, x_1^4 = \frac{4\pi^2}{6 \cdot 4 \cdot 2} 3 \cdot 1 = \frac{\pi^2}{4}$$

$$\frac{1}{r^4} \int d\Omega \, x_1^2 x_2^2 = \frac{4\pi^2}{6 \cdot 4 \cdot 2} = \frac{\pi^2}{12}$$

$$\mathcal{N}^2 \pi^2 \left(\frac{1}{4} - \frac{1}{6} + \frac{1}{4}\right) = 1$$

$$\mathcal{N} = \frac{\sqrt{6}}{\sqrt{2\pi}}$$

Exercise 3.4

Show that the hyperspherical harmonics with $\lambda = 1$ in Table 2.3 fulfill the sum rule of equation (3.72). Show that they are properly normalized.

Solution

$$\sum_\mu Y_{1,\mu}^*(\mathbf{u}')Y_{1,\mu}(\mathbf{u}) = \frac{1}{\pi^2}[(u_1' - iu_2')(u_1 + iu_2) + 2u_3'u_3$$

$$+(u_1' + iu_2')(u_1 - iu_2) + 2u_4'u_4]$$

$$= \frac{2}{\pi^2}\mathbf{u}' \cdot \mathbf{u} = \frac{1}{\pi^2}C_1^1(\mathbf{u}' \cdot \mathbf{u})$$

$$= \frac{(2\lambda + d - 2)}{(d-2)I(0)}C_1^1(\mathbf{u}' \cdot \mathbf{u})$$

From equation (3.49),

$$\int d\Omega |Y_{1,1,\pm 1}(\mathbf{u})|^2 = 2\int d\Omega(u_1^2 + u_2^2) = 2\left(\frac{1}{4} + \frac{1}{4}\right) = 1$$

$$\int d\Omega |Y_{1,1,0}(\mathbf{u})|^2 = 4\int d\Omega u_3^2 = \frac{4}{4} = 1$$

$$\int d\Omega |Y_{1,1,\pm 1}(\mathbf{u})|^2 = 4\int d\Omega u_4^2 = \frac{4}{4} = 1$$

Exercise 3.5

Use equation (3.72) to show that

$$\frac{2\lambda + d - 2}{(d-2)I(0)}\int d\Omega\ C_\lambda^\alpha(\hat{\mathbf{x}} \cdot \hat{\mathbf{x}}')C_{\lambda'}^\alpha(\hat{\mathbf{x}} \cdot \hat{\mathbf{x}}'') = \delta_{\lambda'\lambda}C_\lambda^\alpha(\hat{\mathbf{x}}' \cdot \hat{\mathbf{x}}'')$$

Solution

$$\frac{2\lambda + d - 2}{(d-2)I(0)}\int d\Omega\ C_\lambda^\alpha(\hat{\mathbf{x}} \cdot \hat{\mathbf{x}}')C_{\lambda'}^\alpha(\hat{\mathbf{x}} \cdot \hat{\mathbf{x}}'')$$

$$= \frac{(d-2)I(0)}{2\lambda + d - 2}\sum_{\{\mu\}}\sum_{\{\mu'\}}Y_{\lambda,\{\mu\}}^*(\hat{\mathbf{x}}')Y_{\lambda',\{\mu'\}}(\hat{\mathbf{x}}'')\int d\Omega Y_{\lambda',\{\mu'\}}^*(\hat{\mathbf{x}})Y_{\lambda,\{\mu\}}(\hat{\mathbf{x}})$$

$$= \frac{(d-2)I(0)}{2\lambda + d - 2}\sum_{\{\mu\}}\sum_{\{\mu'\}}Y_{\lambda,\{\mu\}}^*(\hat{\mathbf{x}}')Y_{\lambda',\{\mu'\}}(\hat{\mathbf{x}}')\delta_{\lambda'\lambda}\delta_{\{\mu'\},\{\mu\}}$$

$$= \delta_{\lambda'\lambda}\frac{(d-2)I(0)}{2\lambda + d - 2}\sum_{\{\mu\}}Y_{\lambda,\{\mu\}}^*(\hat{\mathbf{x}}')Y_{\lambda',\{\mu'\}}(\hat{\mathbf{x}}'')$$

$$= \delta_{\lambda'\lambda}C_\lambda^\alpha(\hat{\mathbf{x}}' \cdot \hat{\mathbf{x}}'')$$

Exercise 4.1

Use equations (4.17)-(4.20) to derive equation (4.21).

Solution

$$\frac{d\Omega}{d^3p} = \frac{\sin^2\chi \, \sin\theta_p \, d\chi d\theta_p d\phi_p}{p^2 dp \, \sin\theta_p d\theta_p d\phi_p}$$

$$d\Omega = \left(\frac{\sin^2\chi}{p^2}\right)\left(\frac{d\chi}{dp}\right) d^3p$$

$$\frac{d\chi}{dp} = \frac{2k_\mu}{k_\mu^2 + p^2} = \frac{\sin\chi}{p}$$

$$d\Omega = \left(\frac{2k_\mu}{k_\mu^2 + p^2}\right)^3 d^3p$$

Exercise 4.2

Use equation (4.39) to generate the associated Laguerre polynomials of Table 4.2.

Solution

$$L_p^0 = e^\rho \frac{d^p}{d\rho^p}\left(e^{-\rho}\rho^p\right)$$

$$L_0^0 = \rho^0 = 1$$

$$L_1^0 = e^\rho \frac{d}{d\rho}\left(e^{-\rho}\rho\right) = 1 - \rho$$

$$L_2^0 = e^\rho \frac{d^2}{d\rho^2}\left(e^{-\rho}\rho^2\right) = 2 - 4\rho + \rho^2$$

$$L_3^0 = e^\rho \frac{d^3}{d\rho^3}\left(e^{-\rho}\rho^3\right) = 6 - 18\rho + 9\rho^2 - \rho^3$$

$$L_1^1 = \frac{d}{d\rho}(1 - \rho) = -1$$

$$L_2^1 = \frac{d}{d\rho}\left(2 - 4\rho + \rho^2\right) = -4 + 2\rho$$

$$L_2^2 = \frac{d}{d\rho}(-4+2\rho) = 2$$

$$L_3^1 = \frac{d}{d\rho}\left(6 - 18\rho + 9\rho^2 - \rho^3\right) = -18 + 18\rho - 3\rho^2$$

$$L_3^2 = \frac{d}{d\rho}\left(-18 + 18\rho - 3\rho^2\right) = 18\rho - 6\rho$$

$$L_3^3 = \frac{d}{d\rho}(18\rho - 6\rho) = -6$$

Exercise 4.3

From equation (4.38) and the associated Laguerre polynomials in Table 4.2, calculate the parabolic hydrogenlike orbitals shown in Table 4.3. Express these functions as linear combinations of $\chi_{nlm}(\mathbf{x})$.

Solution

Let $t \equiv k_\mu r$. Then, since $\xi = r + z$ and $\eta = r - z$

$$w_{0,0,0} = \left(\frac{k_\mu^3}{\pi}\right)^{1/2} e^{-k_\mu(\xi+\eta)/2} L_0^0(k_\mu\xi) L_0^0(k_\mu\eta)$$

$$= \left(\frac{k_\mu^3}{\pi}\right)^{1/2} e^{-k_\mu(\xi+\eta)/2}$$

$$= \left(\frac{k_\mu^3}{\pi}\right)^{1/2} e^{-t}$$

$$w_{1,0,0} = \left(\frac{k_\mu^3}{\pi}\right)^{1/2} e^{-k_\mu(\xi+\eta)/2} L_1^0(k_\mu\xi) L_0^0(k_\mu\eta)$$

$$= \left(\frac{k_\mu^3}{\pi}\right)^{1/2} e^{-k_\mu(\xi+\eta)/2}(1 - k_\mu\xi)$$

$$= \left(\frac{k_\mu^3}{\pi}\right)^{1/2} e^{-t}(1 - t - t_3)$$

$$w_{0,1,0} = \left(\frac{k_\mu^3}{\pi}\right)^{1/2} e^{-k_\mu(\xi+\eta)/2} L_0^0(k_\mu\xi) L_1^0(k_\mu\eta)$$

$$= \left(\frac{k_\mu^3}{\pi}\right)^{1/2} e^{-k_\mu(\xi+\eta)/2}(1 - k_\mu\eta)$$

$$= \left(\frac{k_\mu^3}{\pi}\right)^{1/2} e^{-t}(1 - t + t_3)$$

$$w_{0,0,\pm 1} = \left(\frac{k_\mu^5}{\pi}\right)^{1/2} e^{-k_\mu(\xi+\eta)/2}(\xi\eta)^{1/2} e^{\pm i\phi} L_1^1(k_\mu\xi) L_1^1(k_\mu\eta)$$

$$= \left(\frac{k_\mu^5}{\pi}\right)^{1/2} e^{-k_\mu(\xi+\eta)/2}(\xi\eta)^{1/2} e^{\pm i\phi}$$

$$= \left(\frac{k_\mu^3}{\pi}\right)^{1/2} e^{-t}(t_1 \pm it_2)$$

$$
\begin{aligned}
w_{0,0,0} &= \chi_{1,0,0} \\
w_{1,0,0} &= \frac{1}{\sqrt{2}}(\chi_{2,0,0} - \chi_{2,1,0}) \\
w_{0,1,0} &= \frac{1}{\sqrt{2}}(\chi_{2,0,0} + \chi_{2,1,0}) \\
w_{0,0,1} &= -\chi_{2,1,1} \\
w_{0,0,-1} &= \chi_{2,1,-1}
\end{aligned}
$$

Exercise 5.1

Calculate the Shibuya-Wulfman integrals $S_{1,0,0}^{1,0,0}$, $S_{1,0,0}^{1,0,0}$, and $S_{1,0,0}^{2,1,0}$ in terms of the universal function $\mathcal{F}_{k,l}(s)$ by means of equation (5.53). Compare your results with those shown in Table 5.1. Is the integral $S_{1,0,0}^{2,1,0}$ real?

Solution

$$\frac{1}{2\pi^2}\int d\Omega\, e^{i\mathbf{p}\cdot\mathbf{R}}\, u_4^k h_l(u_j) = \mathcal{F}_{k,l}(s)h_l(s_j)$$

$$S_{1,0,0}^{1,0,0} = \int d\Omega\, e^{i\mathbf{p}\cdot\mathbf{R}}\, Y_{0,0,0}^*(\mathbf{u})Y_{0,0,0}(\mathbf{u})$$

$$= \frac{1}{2\pi^2} \int d\Omega \ e^{i\mathbf{p}\cdot\mathbf{R}}$$

$$= \mathcal{F}_{0,0}(s) = (1+s)e^{-s}$$

$$S_{1,0,0}^{2,0,0} = \int d\Omega \ e^{i\mathbf{p}\cdot\mathbf{R}} \ Y_{0,0,0}^*(\mathbf{u})Y_{1,0,0}(\mathbf{u})$$

$$= \frac{2}{2\pi^2} \int d\Omega \ e^{i\mathbf{p}\cdot\mathbf{R}}u_4$$

$$= \mathcal{F}_{1,0}(s) = -\frac{2}{3}s^2 e^{-s}$$

$$S_{1,0,0}^{2,1,0} = \int d\Omega \ e^{i\mathbf{p}\cdot\mathbf{R}} \ Y_{0,0,0}^*(\mathbf{u})Y_{1,1,0}(\mathbf{u})$$

$$= \frac{-2i}{2\pi^2} \int d\Omega \ e^{i\mathbf{p}\cdot\mathbf{R}}u_3$$

$$= -2i\mathcal{F}_{0,1}(s)s_3 = \frac{2}{3}(1+s)e^{-s}s_3$$

The integral is real.

Exercise 5.2

Calculate

$$S_{1,0,0}^{2,1,0} = \int d\Omega \ e^{i\mathbf{p}\cdot\mathbf{R}}Y_{0,0,0}^*(\mathbf{u})Y_{1,1,0}(\mathbf{u})$$

by means of equations (5.40)-(5.42). Show that the result agrees with Table 5.1 and Exercise 1.

Solution

$$S_{1,0,0}^{2,1,0} = \int d\Omega \ e^{i\mathbf{p}\cdot\mathbf{R}} \ Y_{0,0,0}^*(\mathbf{u})Y_{1,1,0}(\mathbf{u})$$

$$= \frac{-2i}{2\pi^2} \int d\Omega \ e^{i\mathbf{p}\cdot\mathbf{R}}u_3$$

$$\int d\Omega e^{i\mathbf{p}\cdot\mathbf{R}} \ Y_{\mu'}^*(\mathbf{u})Y_\mu(\mathbf{u}) = \left(\frac{2\pi}{k_{\mu'}}\right)^{3/2} \sum_{\mu''} \chi_{\mu''}(\mathbf{R})c_{\mu',\mu}^{\mu''}$$

$$c_{\mu',\mu}^{\mu''} = \int d\Omega \ (1+u_4)Y_{\mu''}^*(\mathbf{u})Y_{\mu'}^*(\mathbf{u})Y_\mu(\mathbf{u})$$

$$c_{0,0,0;1,1,0}^{\mu''} = \frac{-2i}{2\pi^2} \int d\Omega \ (u_3 + u_3 u_4)Y_{\mu''}^*(\mathbf{u})$$

$$c_{0,0,0;1,1,0}^{1,1,0} = \frac{(-2i)(2i)}{(\sqrt{2\pi})^3} \int d\Omega \, (u_3 + u_3 u_4) u_3 = \frac{1}{\sqrt{2\pi}}$$

$$c_{0,0,0;1,1,0}^{2,1,0} = \frac{(-2i)(-2i\sqrt{6})}{(\sqrt{2\pi})^3} \int d\Omega \, (u_3 + u_3 u_4) u_3 u_4 = -\frac{1}{\sqrt{6}} \frac{1}{\sqrt{2\pi}}$$

$$S_{1,0,0}^{2,1,0} = \frac{1}{\sqrt{2\pi}} \left(\frac{2\pi}{k_\mu}\right)^{3/2} \left[\chi_{2,1,0}(\mathbf{R}) - \frac{1}{\sqrt{6}} \chi_{3,1,0}(\mathbf{R})\right]$$

$$= \frac{2}{3}(1+s)e^{-s}$$

Exercise 5.3

Use (3.49) to write down an angular integration theorem analogous to (5.42) for the case where $d = 3$.

Solution

$$r^{-n} \int d\Omega \prod_{j=1}^{3} x_j^{n_j} = \frac{2\pi^{3/2}}{\Gamma(\frac{3}{2})(n+1)!!} \prod_{j=1}^{3} (n_j - 1)!!$$

$$= \frac{4\pi}{(n+1)!!} \prod_{j=1}^{3} (n_j - 1)!!$$

Exercise 6.1

Starting with equations (6.44) and (6.46), derive equations (6.47) and (6.48).

Solution

$$K_{\tau',\tau}^{2} = -\frac{1}{k_\mu} \sqrt{\frac{Z_{a'} Z_a}{n'n}} \int d^3x \chi_{\mu'}^{*}(\mathbf{x} - \mathbf{X}_{a'}) v(\mathbf{x}) \chi_\mu(\mathbf{x} - \mathbf{X}_a)$$

where $\tau \equiv \{a, \mu\} = \{a, n, l, m\}$ and $a = 1, 2$

$$v(\mathbf{x}) = -\frac{Z_1}{|\mathbf{x} - \mathbf{X}_1|} - \frac{Z_2}{|\mathbf{x} - \mathbf{X}_2|}$$

$$K^2_{1,\mu',1,\mu} = \frac{1}{k_\mu}\frac{Z_1}{\sqrt{n'n}}\int d^3x\chi^*_{\mu'}(\mathbf{x}-\mathbf{X}_1)\chi_\mu(\mathbf{x}-\mathbf{X}_1)\left(\frac{Z_1}{|\mathbf{x}-\mathbf{X}_1|}+\frac{Z_2}{|\mathbf{x}-\mathbf{X}_2|}\right)$$

$$= \frac{1}{k_\mu}\frac{Z_1^2}{\sqrt{n'n}}\int d^3x\chi^*_{\mu'}(\mathbf{x})\chi_\mu(\mathbf{x})\frac{1}{r}$$

$$+\frac{1}{k_\mu}\frac{Z_1Z_2}{\sqrt{n'n}}\int d^3x\chi^*_{\mu'}(\mathbf{x})\chi_\mu(\mathbf{x})\frac{1}{|\mathbf{x}-\mathbf{R}|}$$

where $\mathbf{R} \equiv \mathbf{X}_2 - \mathbf{X}_1$

$$K^2_{1,\mu',1,\mu} = \frac{Z_1^2}{n^2}\delta_{\mu',\mu}+\frac{1}{k_\mu}\frac{Z_1Z_2}{\sqrt{n'n}}\int d^3x\chi^*_{\mu'}(\mathbf{x})\chi_\mu(\mathbf{x})\frac{1}{|\mathbf{x}-\mathbf{R}|}$$

and similarly

$$K^2_{2,\mu',2,\mu} = \frac{Z_2^2}{n^2}\delta_{\mu',\mu}+\frac{1}{k_\mu}\frac{Z_1Z_2}{\sqrt{n'n}}\int d^3x\chi^*_{\mu'}(\mathbf{x})\chi_\mu(\mathbf{x})\frac{1}{|\mathbf{x}+\mathbf{R}|}$$

Exercise 6.2

Show that

$$\frac{1}{k_\mu}\int d^3x\ \chi^*_{1s}(\mathbf{x})\chi_{1s}(\mathbf{x})\frac{1}{|\mathbf{x}-\mathbf{R}|} = \frac{1}{s}-\frac{e^{-2s}}{s}(1+s)$$

(Table 6.1) where $s = k_\mu R$, and where

$$\chi_{1s}(\mathbf{x}) = \left(\frac{k_\mu^3}{\pi}\right)^{1/2}e^{-k_\mu r}$$

What is the limit of the integral as $s \to 0$?

Solution

Let $\mathbf{t} = k_\mu\mathbf{x}$, $\mathbf{s} = k_\mu\mathbf{R}$, and

$$t_> \equiv \begin{cases} s \text{ if } s > t \\ t \text{ if } s < t \end{cases}$$

Then

$$\frac{1}{k_\mu} \int d^3x \, \chi_{1s}^*(\mathbf{x})\chi_{1s}(\mathbf{x})\frac{1}{|\mathbf{x} - \mathbf{R}|} = \frac{1}{k_\mu}\left(\frac{k_\mu^3}{\pi}\right)\int d^3x \, e^{-2k_\mu r}\frac{1}{|\mathbf{x} - \mathbf{R}|}$$

$$= \frac{1}{\pi}\int d^3t \, e^{-2t}\frac{1}{|\mathbf{t} - \mathbf{s}|}$$

$$= \frac{1}{\pi}\int d^3t \, e^{-2t}\sum_{l=0}^{\infty}\frac{t_<^l}{t_>^{l+1}}P_l(\hat{\mathbf{t}}\cdot\hat{\mathbf{s}})$$

$$= \frac{4}{s}\int_0^s dt \, t^2 \, e^{-2t} + 4\int_s^\infty dt \, t \, e^{-2t}$$

$$= \frac{1}{s} - \frac{e^{-2s}}{s}(1 + s)$$

$$\lim_{s \to 0}\left[\frac{1}{s} - \frac{e^{-2s}}{s}(1 + s)\right] = \lim_{s \to 0}\left[\frac{1}{s} - \frac{(1 - 2s)}{s}(1 + s)\right] = 1$$

Exercise 6.3

Consider two nuclei with charges Z_1 and Z_2. Write down the matrices $K_{\tau',\tau}$ and $K_{\tau',\tau}^2$ for the case where the basis set consists of a single $1s$ atomic orbital localized on each center. Why is the square of the first matrix not equal to the second?

Solution

$$K^2 = \begin{pmatrix} Z_1^2 + \frac{Z_1 Z_2}{s}[1 - e^{-2s}(1 + s)] & \sqrt{Z_1 Z_2}(Z_1 + Z_2)e^{-s}(1 + s) \\ \sqrt{Z_1 Z_2}(Z_1 + Z_2)e^{-s}(1 + s) & Z_2^2 + \frac{Z_1 Z_2}{s}[1 - e^{-2s}(1 + s)] \end{pmatrix}$$

$$K = \begin{pmatrix} Z_1 & \sqrt{Z_1 Z_2}e^{-s}(1 + s) \\ \sqrt{Z_1 Z_2}e^{-s}(1 + s) & Z_2 \end{pmatrix}$$

$$(K)^2 = \begin{pmatrix} Z_1^2 + Z_1 Z_2 e^{-2s}(1 + s)^2 & \sqrt{Z_1 Z_2}(Z_1 + Z_2)e^{-s}(1 + s) \\ \sqrt{Z_1 Z_2}(Z_1 + Z_2)e^{-s}(1 + s) & Z_2^2 + Z_1 Z_2 e^{-2s}(1 + s)^2 \end{pmatrix}$$

The inequality $K^2 \neq (K)^2$ is due to the fact that the basis set is truncated, so that the sum

$$\sum_{\tau''} K_{\tau,\tau''} K_{\tau'',\tau'}$$

does not run over all possible values of τ''.

Exercise 7.1

Calculate $\mathcal{G}_{0,0}(s)$, $\mathcal{G}_{1,0}(s)$, and $\mathcal{G}_{0,1}(s)$, using Table 5.2 and the definition

$$\mathcal{G}_{k,l}(s) \equiv \mathcal{F}_{k,l}(s) + \mathcal{F}_{k+1,l}(s)$$

Solution

$$
\begin{aligned}
\mathcal{G}_{0,0}(s) &= \mathcal{F}_{0,0}(s) + \mathcal{F}_{1,0}(s) \\
&= (1+s)e^{-s} + \frac{1}{3}s^2 e^{-s} \\
&= (1 + s + \frac{1}{3}s^2)e^{-s}
\end{aligned}
$$

$$
\begin{aligned}
\mathcal{G}_{1,0}(s) &= \mathcal{F}_{1,0}(s) + \mathcal{F}_{2,0}(s) \\
&= -\frac{1}{3}s^2 e^{-s} - \frac{1}{12}(3 + 3s - 2s^2 + s^3)e^{-s} \\
&= -\frac{1}{12}(3 + 3s + 2s^2 + s^3)e^{-s}
\end{aligned}
$$

$$
\begin{aligned}
\mathcal{G}_{0,1}(s) &= \mathcal{F}_{0,1}(s) + \mathcal{F}_{1,1}(s) \\
&= \frac{i}{3}(1+s)e^{-s} - \frac{i}{12}(1 + s - s^2)e^{-s} \\
&= \frac{i}{12}(3 + 3s + s^2)e^{-s}
\end{aligned}
$$

Exercise 7.2

Use the results of Exercise 7.1 to calculate the Sturmian overlap integral

$$I = \int d^3x_j \; \chi_{1s}^*(\mathbf{x}_j)\chi_{1s}(\mathbf{x}_j + \mathbf{R})$$

Solution

$$
\begin{aligned}
I &= \int d^3p_j \; e^{i\mathbf{p}_j \cdot \mathbf{R}} \chi_{1s}^{*t}(\mathbf{p}_j)\chi_{1s}^t(\mathbf{p}_j) \\
&= \int d\Omega \; e^{i\mathbf{p}_j \cdot \mathbf{R}}(1 + u_4)Y_{0,0,0}^*(\mathbf{u})Y_{0,0,0}(\mathbf{u}) \\
&= \frac{1}{2\pi^2} \int d\Omega \; e^{i\mathbf{p}_j \cdot \mathbf{R}}(1 + u_4) \\
&= \mathcal{G}_{0,0}(s) = (1 + s + \frac{1}{3}s^2)e^{-s}
\end{aligned}
$$

where $s = k_\mu R$.

Exercise 7.3

Calculate the integral I of Exercise 7.2 using the ellipsoidal coordinates $\xi = (r_a + r_b)/R$, $\eta = (r_a - r_b)/R$ and ϕ where $r_a^2 = \mathbf{x}_j \cdot \mathbf{x}_j$ and $r_b^2 = (\mathbf{x}_j + \mathbf{R}) \cdot (\mathbf{x}_j + \mathbf{R})$ and where ϕ has its usual meaning. In ellipsoidal coordinates, the volume element is given by

$$d^3x = \frac{R^3}{8}\left(\xi^2 - \eta^2\right)d\xi d\eta d\phi$$

Compare your answer with the results of Exercise 7.2. Could ellipsoidal coordinates be used to calculate Shibuya-Wulfman integrals?

Solution

$$
\begin{aligned}
I &= \frac{k_\mu^3}{\pi} \int d^3x_j \; e^{-k_\mu(r_a + r_b)} \\
&= \frac{s^3}{8\pi} \int_1^\infty d\xi \int_{-1}^1 d\eta \int_0^{2\pi} d\phi \left(\xi^2 - \eta^2\right)e^{-s\xi}
\end{aligned}
$$

$$= \frac{s^3}{4} \int_1^\infty d\xi e^{-s\xi} \int_{-1}^1 d\eta \left(\xi^2 - \eta^2 \right)$$

$$= \frac{s^3}{4} \int_1^\infty d\xi e^{-s\xi} \left(2\xi^2 - \frac{2}{3} \right)$$

$$= (1 + s + \frac{1}{3}s^2)e^{-s}$$

Ellipsoidal coordinates also offer an alternative method for evaluating Shibuya-Wulfman integrals.

Exercise 8.1

Show that for $j = 1/2$, $l = 0$, and $M = 1/2$, the 4-component solution to the hydrogenlike Dirac equation can be written in the form:

$$\chi_{n,j,l,M}(\mathbf{x}) = \chi_{n,\frac{1}{2},0,\frac{1}{2}}(\mathbf{x}) = \begin{pmatrix} ig_{n,\frac{1}{2},0}(r)Y_{0,0}(\theta,\phi) \\ 0 \\ \sqrt{\frac{1}{3}}f_{n,\frac{1}{2},0}(r)Y_{1,0}(\theta,\phi) \\ -\sqrt{\frac{2}{3}}f_{n,\frac{1}{2},0}(r)Y_{1,1}(\theta,\phi) \end{pmatrix}$$

What is the form of the solution corresponding to $j = 1/2$, $l = 0$, and $M = -1/2$?

Solution

From (8.8) we have:

$$\Omega_{j,l,M} = \Omega_{\frac{1}{2},0,\frac{1}{2}} = \begin{pmatrix} Y_{0,0} \\ 0 \end{pmatrix}$$

while from (8.9) with $\bar{l} = 2j - l = 1$,

$$\Omega_{\frac{1}{2},1,\frac{1}{2}} = \begin{pmatrix} -\sqrt{\frac{1}{3}}Y_{1,0} \\ \sqrt{\frac{2}{3}}Y_{1,1} \end{pmatrix}$$

Therefore (8.5) yields

$$\chi_{n,\frac{1}{2},0,\frac{1}{2}}(\mathbf{x}) = \begin{pmatrix} ig_{n,\frac{1}{2},0}(r)Y_{0,0}(\theta,\phi) \\ 0 \\ \sqrt{\frac{1}{3}}f_{n,\frac{1}{2},0}(r)Y_{1,0}(\theta,\phi) \\ -\sqrt{\frac{2}{3}}f_{n,\frac{1}{2},0}(r)Y_{1,1}(\theta,\phi) \end{pmatrix}$$

Similarly, when $M = -1/2$

$$\chi_{n,\frac{1}{2},0,-\frac{1}{2}}(\mathbf{x}) = \begin{pmatrix} 0 \\ ig_{n,\frac{1}{2},0}(r)Y_{0,0}(\theta,\phi) \\ \sqrt{\frac{2}{3}}f_{n,\frac{1}{2},0}(r)Y_{1,-1}(\theta,\phi) \\ -\sqrt{\frac{1}{3}}f_{n,\frac{1}{2},0}(r)Y_{1,0}(\theta,\phi) \end{pmatrix}$$

Exercise 8.2

Letting $b_\mu = 1$, find the values of κ, n_r, γ, N, and ϵ_μ for the solutions to (8.1) in the $n = 1$ and $n = 2$ shells.

Solution

From equations (8.12)-(8.15) and (8.19) we obtain:

$\{n, j, l\}$	κ	n_r	γ	N	$\epsilon_\mu/(mc^2)$
$\{1, 1/2, 0\}$	-1	0	0.9999733766	1.00000000000	0.99997337665
$\{2, 1/2, 0\}$	-1	1	0.9999733766	1.99998668828	0.99999334414
$\{2, 1/2, 1\}$	1	1	0.9999733766	1.99998668828	0.99999334414
$\{2, 3/2, 1\}$	-2	0	1.9999866884	2.00000000000	0.99999334423

Exercise 8.3

The energies ϵ_μ calculated in Exercise 8.2 include the electron rest energy mc^2 and are expressed in units of mc^2. Subtract the rest energy from the calculated values, and express the results in Hartrees.

Solution

In order to convert from units of mc^2 to Hartrees, we must multiply by $(\hbar c/e^2)^2 = (137.0429)^2$. The resulting energies

$$E_\mu = \left(\frac{\hbar c}{e^2}\right)^2 \left(\epsilon_\mu - mc^2\right)$$

are

$\{n, j, l\}$	E_μ
$\{1, 1/2, 0\}$	-0.50000665592
$\{2, 1/2, 0\}$	-0.12500207998
$\{2, 1/2, 1\}$	-0.12500207998
$\{2, 3/2, 1\}$	-0.12500041599

Bibliography

1. Ahlberg, R. and Lindner, P., *The Fermi correlation for electrons in momentum space*, J Phys B, Vol 9 (17), p 2963-9, 1976.

2. Ahlenius, Tor and Lindner, Peter., *Semiempirical MO wave functions in momentum space*, J Phys B, Vol 8 (5), p 778-95, 1975.

3. Akhiezer, A.I. and Berestetskii, V.B., **Quantum Electrodynamics**, Interscience, New York, 1965.

4. Allan, Neil L. and Cooper, David L., *Local density approximations and momentum-space properties in light molecules and ionic solids*, J Chem Soc, Faraday Trans 2, Vol 83 (9), p 1675-87, 1987.

5. Allan, Neil L. and Cooper, David L., *Momentum space properties and local density approximations in small molecules: a critical appraisal*, J Chem Phys, Vol 84 (10), p 5594-605, 1986.

6. Alliluev, S.P., Sov Phys JETP, Vol 6, p 156, 1958.

7. Amiet, J.-P. et Huguenin, P., *Mécaniques classique et quantiques dans l'espace de phase*, Université de Neuchâtel, 1981.

8. Amos, A.T. and Hall, G.G., Proc. Roy. Soc. London, Vol A263, p 483, 1961.

9. Anderson, R.W.; Aquilanti, V.; Cavalli, S. and Grossi, G., J Phys Chem, Vol 95, p 8184, 1991.

10. Anderson, R.W.; Aquilanti, V.; Cavalli, S. and Grossi, G., J Phys Chem, Vol 97, p 2443, 1993.

11. Aquilanti, V. and Cavalli, S., *Coordinates for molecular dynamics*, J Chem Phys, Vol 85, p 1355-1361, 1986.

12. Aquilanti, V., Cavalli, S., De Fazio, D, and Grossi, G. *Hyperangular Momentum: Applications to Atomic and Molecular Science*, in **New Methods in Quantum Theory**, Tsipis, C.A., Popov, V.S., Herschbach, D.R., and Avery, J.S., Eds., Kluwer, Dordrecht, 1996.

13. Aquilanti, V.; Cavalli, S. and Grossi, G., *Hyperspherical coordinates for molecular dynamics by the method of trees and the mapping of potential-energy surfaces for triatomic systems*, J Chem Phys, Vol 85, p 1362, 1986.

14. Aquilanti, V.; Grossi, G.; Laganá, A.; Pelikan, E., and Klar, H., *A decoupling scheme for a 3-body problem treated by expansionns into hyperspherical harmonics. The hydrogen molecular ion*, Lett Nuovo Cimento, Vol 41, 541, 1984.

15. Aquilanti, V.; Grossi, G., and Laganá, A., *On hyperspherical mapping and harmonic expansions for potential energy surfaces* J Chem Phys, Vol 76, p 1587-1588, 1982.

16. Aquilanti, V.; Laganá, A., and Levine, R.D., Chem Phys Lett, Vol 158, p 87, 1989.

17. Aquilanti, V. and Cavalli, S., Chem Phys Lett, Vol 141, p 309, 1987.

18. Aquilanti, V.; Cavalli, S.; Grossi, G.; Rosi, M.; Pellizzari, V.; Sgamellotti, A., and Tarantelli, F.; Chem Phys Lett, Vol 16, p 179, 1989.

19. Aquilanti, V.; Cavalli, S.; Grossi, G., and Anderson, R.W., J Chem Soc Faraday Trans, Vol 86, p 1681, 1990.

20. Aquilanti, V.; Benevente, L.; Grossi, G. and Vecchiocattivi, F., *Coupling schemes for atom-diatom interactions, and an adiabatic decoupling treatment of rotational temperature effects on glory scattering*, J Chem Phys, Vol 89, 751-761, 1988.

21. Aquilanti, V. and Grossi, G., *Angular momentum coupling schemes in the quantum mechanical treatment of P-state atom collisions* J Chem Phys, Vol 73, p 1165-1172, 1980.

22. Aquilanti, V.; Cavalli, S., and Grossi, G., Theor Chem Acta, Vol 79, p 283, 1991.

23. Aquilanti, V. and Cavalli, S., **Few Body Systems**, Suppl 6, p 573, 1992.

24. Aquilanti, V. and Grossi, G., Lett Nuovo Cimento, Vol 42, p 157, 1985.

25. Aquilanti, V. Cavalli, S. and De Fazio, D., *Angular and hyperangular momentum coupling coefficients as Hahn polynomials*, J Phys Chem Vol 99, p 15694, 1995.

26. Aquilanti, V., Cavalli, S., Coletti, C. and Grossi, G., *Alternative Sturmian bases and momentum space orbitals; an application to the hydrogen molecular ion*, Chem Phys Vol 209, p 405, 1996.

27. Aquilanti, V., Cavalli, S. and Coletti, C., *The d-dimensional hydrogen atom; hyperspherical harmonics as momentum space orbitals and alternative Sturmian basis sets*, Chem Phys Vol 214, p 1, 1997.

28. Aquilanti, V., and Avery, J., *Generalized Potential Harmonics and Contracted Sturmians*, Chem. Phys. Letters, Vol 267, p 1, 1997.

29. Avery, John and Ørmen, Per-Johan, Int. J. Quantum Chem. Vol 18, p 953, 1980.

30. Avery, John, **Hyperspherical Harmonics; Applications in Quantum Theory**, Kluwer Academic Publishers, Dordrecht, 1989.

31. Avery, John, *Hyperspherical Sturmian Basis Functions in Reciprocal Space*, in **New Methods in Quantum Theory**, Tsipis, C.A., Popov, V.S., Herschbach, D.R., and Avery, J.S., Eds., Kluwer, Dordrecht, 1996.

32. Avery, John and Antonsen, Frank, *A new approach to the quantum mechanics of atoms and small molecules*, Int J Quantum Chem, Symposium 23, p 159, 1989.

33. Avery, John and Antonsen, Frank, *Iteration of the Schrödinger equation, starting with Hartree-Fock wave functions*, Int J Quantum Chem, Vol 42, p 87, 1992.

34. Avery, John and Antonsen, Frank, Theor. Chim. Acta, Vol 85, p 33, 1993.

35. Avery, John and Herschbach, Dudley R., *Hyperspherical sturmian basis functions*, Int J Quantum Chem, Vol 41, p 673, 1992.

36. Avery, John and Wen, Zhen-Yi, *A Formulation of the quantum mechanical many-body in terms of hyperspherical coordinates*, Int J Quantum Chem Vol 25, p 1069, 1984.

37. Avery, John, *Correlation in iterated solutions of the momentum-space Schrödinger equation*, Chem Phys Lett, Vol 138 (6), p 520-4, 1987.

38. Avery, John, *Hyperspherical Harmonics; Some Properties and Applications*, in **Conceptual Trends in Quantum Chemistry**, Kryachko, E.S., and Calais, J.L., Eds, Kluwer, Dordrecht, 1994.

39. Avery, John, Hansen, T.B., Wang, M. and Antonsen, F., *Sturmian basis sets in momentum space*, Int J Quant Chem Vol 57, p 401, 1996.

40. Avery, John, and Hansen, Tom Børsen, *A momentum-space picture of the chemical bond* Int J Quant Chem Vol 60, p 201, 1996.

41. Avery, John, *Many-Particle Sturmians*, J Math Chem, Vol 21, p 285, 1997.

42. Avery, John and Antonsen, Frank, *Relativistic Sturmian Basis Functions*, J. Math. Chem. Vol 24, p175, 1998.

43. Avery, John, *A Formula for Angular and Hyperangular Integration*, J. Math. Chem., Vol 24, p 169, 1998.

44. Avery, John, *Many-electron Sturmians applied to atoms and ions*, J. Mol. Struct. Vol 458, p 1, 1999.

45. Avery, John, *Many-Electron Sturmians as an Alternative to the SCF-CI Method*, Adv. Quantum Chem., Vol. 31, p 201, 1999.

46. Ballot, L. and Farbre de la Ripelle, M., *Application of the hyperspherical formalism to trinucleon bound-state problems*, Ann Phys, Vol. 127, p 62, 1980.

47. Bandar, M. and Itzyksen, C., *Group theory and the H atom*, Rev Mod Phys Vol 38, p 330, p 346, 1966.

48. Bang, J.M. and Vaagen, J.S., *The Sturmian expansion: a well-depth-method for orbitals in a deformed potential*, Z Phys A, Vol 297 (3), p 223-36, 1980.

49. Bang, J.M., Gareev, F.G., Pinkston, W.T. and Vaagen, J.S., Phys Rep Vol 125, p 253-399, 1985.

50. Bar-yudin, L. E. and Tel-nov, D. A., *Sturmian expansion of the electron density deformation for 3d-metal ions in electric field*, Vestn Leningr Univ, Ser 4: Fiz, Khim (1), p 83-6, 1991.

51. Baretty, Reinaldo; Ishikawa, Yasuyuki; and Nieves, Jose F., *Momentum space approach to relativistic atomic structure calculations*, Int J Quantum Chem, Quantum Chem Symp, Vol 20, p 109-17, 1986.

52. Benesch, Robert and Smith, Vedene H. Jr., *Natural orbitals in momentum space and correlated radial momentum distributions. I. The 1S ground state of Li+*, Int J Quantum Chem, Symp, Vol No. 4, p 131-8, 1971.

53. Biedenharn, L.C. and Louck, J.D., **Angular Momentum in Quantum Physics**, Addison Wesley, Reading, Mass, 1981.

54. Biedenharn, L.C. and Louck, J.D., **The Racah-Wigner Algebra in Quantum Theory**, Addison Wesley, Reading, Mass, 1981.

55. Blinder, S.M., *On Green's functions, propagators, and Sturmians for the nonrelativistic Coulomb problem*, Int J Quantum Chem, Quantum Chem Symp, Vol 18, p 293-307, 1984.

56. Bransden, B.H.; Noble, C.J.; and Hewitt, R.N., *On the reduction of momentum space scattering equations to Fredholm form*, J Phys B: At, Mol Opt Phys, Vol 26 (16), p 2487-99, 1993.

57. Brion, C.E., *Looking at orbitals in the laboratory: the experimental investigation of molecular wave functions and binding energies by electron momentum spectroscopy*, Int J Quantum Chem, Vol 29 (5), p 1397-428, 1986.

58. Brink, D.M. and Satchler, G.R., **Angular Momentum**, Oxford University Press, 1968.

59. Calais, J-L.; Defranceschi, M.; Fripiat, J.G. and Delhalle, J., *Momentum space functions for polymers*, J Phys: Condens Matter, Vol 4 (26), p 5675-91, 1992.

60. Calais, Jean-Louis, *Fukutome classes in momentum space*, Theor Chim Acta, Vol 86 (1-2), p 137-47, 1993.

61. Calais, Jean-Louis, *Orthogonalization in momentum space*, Int J Quantum Chem, Vol 35 (6), p 735-43, 1989.

62. Calais, Jean-Louis, *Pathology of the Hartree-Fock method in configuration and momentum space*, J Chim Phys Phys-Chim Biol, Vol 84 (5), p 601-6, 1987.

63. Chen, Joseph Cheng Yih and Ishihara, Takeshi, *Hydrogenic- and Sturmian-function expansions in three-body atomic problems*, Phys Rev, Vol 186 (1), p 25-38, 1969.

64. Chiu, T.W., *Non-relativistic bound-state problems in momentum space*, J Phys A: Math Gen, Vol 19 (13), p 2537-47, 1986.

65. Cinal, Marek, *Energy functionals in momentum space: exchange energy, quantum corrections, and the Kohn-Sham scheme*, Phys Rev A, Vol 48 (3), p 1893-902, 1993.

66. Clark, Charles W. and Taylor, K.T., *The quadratic Zeeman effect in hydrogen Rydberg series: application of Sturmian functions*, J Phys B, Vol 15 (8), p 1175-93, 1982.

67. Clementi, E., J. Chem. Phys. Vol 38, p 996, 1963.

68. Cohen, L., *Generalized phase-space distribution functions*, J Math Phys, Vol 7, 781-786, 1966.

69. Cohen, Leon and Lee, Chongmoon, *Correlation hole and physical properties: a model calculation*, Int J Quantum Chem, Vol 29 (3), p 407-24, 1986.

70. Coletti, Cecilia, *Struttura Atomica e Moleculare Come Rottura della Simmetria Ipersferica*, Ph.D. thesis, Chemistry Department, University of Perugia, Italy, 1998.

71. Collins, L.A. and Merts, A.L., *Atoms in strong, oscillating electric fields: momentum-space solutions of the time-dependent, three-dimensional Schrödinger equation*, J Opt Soc Am B: Opt Phys, Vol 7 (4), p 647-58, 1990.

72. Coulson, C.A., *Momentum distribution in molecular systems. I. Single bond. III. Bonds of higher order*, Proc Camb Phys Soc, Vol 37, p 55, p 74, 1941.

73. Coulson, C.A. and Duncanson, W.E., *Momentum distribution in molecular systems. II. C and C-H bond*, Proc Camb Phys Soc, Vol 37, p 67, 1941.

74. Dahl, Jens Peder, *The Wigner function*, Physica A, Vol 114, p 439, 1982.

75. Dahl, Jens Peder, *On the group of translations and inversions of phase space and the Wigner function*, Phys Scripta, Vol 25, 499-503, 1982.

76. Dahl, Jens Peder, *Dynamical equations for the Wigner functions*, in **Energy Storage and Redistribution in Molecules**, p 557-571, Ed. J. Hinze, Plenum, New York, 1983.

77. Dahl, Jens Peder, *The phase-space representation of quantum mechanics and the Bohr-Heisenberg correspondence principle*, in

Semiclassical Description of Atomic and Nuclear Collisions, p 379-394, Eds. Bang, J. and De Boer, J., North Holland, Amsterdam, 1985.

78. Dahl, Jens Peder, *The dual nature of phase-space representations*, in **Classical and Quantum Systems**, p 420-423, Eds: Doebner, H.D. and Schroeck, F., Jr., World Scientific, Singapore, 1993.

79. Dahl, Jens Peder, *A phase space essay*, in **Conceptual Trends in Quantum Chemistry**, p 199-224, Eds: Kryachko, E.S. and Calais, J.L., Kluwer Academic Publishers, Dordrecht, Netherlands, 1994.

80. Dahl, Jens Peder and Springborg, Michael, *The Morse oscillator in position space, momentum space, and phase space* J Chem Phys, Vol 88 (7), p 4535-47, 1988.

81. Dahl, Jens Peder and Springborg, Michael, *Wigner's phase-space function and atomic structure. I. The hydrogen atom* , J Mol Phys, Vol 47, p 1001, 1982.

82. Das, G.P.; Ghosh, S.K.; and Sahni, V.C., *On the correlation energy density functional in momentum space*, Solid State Commun, Vol 65 (7), p 719-21, 1988.

83. Davies, R.W. and Davies, K.T.R., *On the Wigner distribution function for an oscillator*, Ann Physics, Vol 89, p 261-273, 1975.

84. De-Prunele, E. *O(4,2) coherent states and hydrogenic atoms*, Phys Rev A, Vol 42 (5), p 2542-9, 1990.

85. De-Windt, Laurent; Defranceschi, Mireille; and Delhalle, Joseph, *Variation-iteration method in momentum space: determination of Hartree-Fock atomic orbitals*, Int J Quantum Chem, Vol 45 (6), p 609-18, 1993.

86. Defranceschi, M.; Suard, M. and Berthier, G., *Numerical solution of Hartree-Fock equations for a polyatomic molecule: linear triatomic hydrogen in momentum space*, Int J Quantum Chem, Vol 25 (5), p 863-7, 1984.

87. Defranceschi, M.; Suard, M.; and Berthier, G., *Epitome of theoretical chemistry in momentum space*, Folia Chim Theor Lat, Vol 18 (2), p 65-82, 1990.

88. Defranceschi, M., *Theoretical investigations of the momentum densities for molecular hydrogen*, Chem Phys, Vol 115 (3), p 349-58, 1987.

89. Defranceschi, Mireille and Delhalle-Joseph., *Numerical solution of the Hartree-Fock equations for quasi-one- dimensional systems: prototypical calculations on the (hydrogen atom) x chain*, Phys Rev B: Condens Matter, Vol 34 (8, Pt. 2), p 5862-73, 1986.

90. Defranceschi, Mireille and Delhalle-Joseph, *Momentum space calculations on the helium atom*, Eur J Phys, Vol 11 (3), p 172-8, 1990.

91. Delande, D. and Gay, J.C., *The hydrogen atom in a magnetic field. Spectrum from the Coulomb dynamical group approach*, J Phys B: At Mol Phys, Vol 19 (6), p L173-L178, 1986.

92. Delhalle, Joseph and Defranceschi and Mireille, *Toward fully numerical evaluation of momentum space Hartree-Fock wave functions. Numerical experiments on the helium atom*, Int J Quantum Chem, Quantum Chem Symp, Vol 21, p 425-33, 1987.

93. Delhalle, Joseph; Fripiat, Joseph G.; and Defranceschi, Mireille, *Improving the one-electron states of ab initio GTO calculations in momentum space. Tests on two-electron systems: hydride, helium, and lithium(1+)*, Bull Soc Chim Belg, Vol 99 (3), p 135-45, 1990.

94. Delhalle, Joseph and Harris, Frank E., *Fourier-representation method for electronic structure of chainlike systems: restricted Hartree-Fock equations and applications to the atomic hydrogen (H)x chain in a basis of Gaussian functions*, Phys Rev B: Condens Matter, Vol 31 (10), p 6755-65, 1985.

95. Deloff, A. and Law, J., *Sturmian expansion method for bound state problems*, Phys Rev C, Vol 21 (5), p 2048-53, 1980.

96. Denteneer, P.J.H. and Van Haeringen, W., *The pseudopotential-density-functional method in momentum space: details and test cases*, J Phys C, Vol 18 (21), p 4127-42, 1985.

97. Desclaux, J.P., Comput. Phys. Commun., Vol 9, p 31, 1975.

98. Desclaux, J.P., Phys. Schripa, Vol 21, p 436, 1980.

99. Dirac, P.A.M., *Note on exchange phenomena in the Thomas atom*, Proc Camb Phil Soc, Vol 26, 376-385, 1930.

100. Dorr, Martin; Potvliege, R.M.; and Shakeshaft, Robin, *Atomic hydrogen irradiated by a strong laser field: Sturmian basis calculations of rates for high-order multiphoton ionization, Raman scattering, and harmonic generation*, J Opt Soc Am B: Opt Phys, Vol 7 (4), p 433-48, 1990.

101. Douglas, Marvin., *Coulomb perturbation calculations in momentum space and application to quantum-electrodynamic hyperfine-structure corrections*, Phys Rev A, Vol 11 (5), p 1527-38, 1975.

102. Drake, G.W.F. and Goldman, S.P., *Relativistic Sturmian and finite basis set methods in atomic physics*, Adv At Mol Phys, Vol 25, p 393-416, 1988.

103. Dube, L.J. and Broad, J.T., *Sturmian discretization. II. The off-shelf Coulomb wavefunction*, J Phys B: At, Mol Opt Phys, Vol 23 (11), p 1711-32, 1990.

104. Dube, Louis J. and Broad, John T., *Sturmian discretization: the off-shell Coulomb wave function*, J Phys B: At, Mol Opt Phys, Vol 22 (18), p L503, 1989.

105. Duchon, C; Dumont-Lepage, M.C.; and Gazeau, J.P., *On two Sturmian alternatives to the LCAO method for a many-center one-electron system*, J Chem Phys, Vol 76 (1), p 445-7, 1982.

106. Duchon, C.; Dumont-Lepage, M.C.; and Gazeau, J.P., *Sturmian methods for the many-fixed-centers Coulomb potential*, J Phys A: Math Gen, Vol 15 (4), p 1227-41, 1982.

107. Duffy, Patrick; Casida, Mark E; Brion, C.E; and Chong, D.P., *Assessment of Gaussian-weighted angular resolution functions in the comparison of quantum-mechanically calculated electron momentum distributions with experiment* Chem Phys, Vol 159 (3), p 347-63, 1992.

108. Duncanson, W.E., *Momentum distribution in molecular systems. IV. H molecule ion, H_2^+*, Proc Camb Phil Soc, Vol 37, p 47, 1941.

109. Dunlap, B.I., Chem Phys Lett, Vol 30, p 39, 1975.

110. Edmonds, A.R., **Angular Momentum in Quantum Chemistry**, Princeton University Press, 1960.

111. Edmonds, A.R., *Quadratic Zeeman effect. I. Application of the sturmian functions*, J Phys B, Vol 6 (8), p 1603-15, 1973.

112. Englefield, M.J., **Group theory and the Coulomb problem**, Wiley-Interscience, New York, 1972.

113. Epstein, P.S., Proc. Natl. Acad. Sci. (USA), Vol 12, p 637, 1926.

114. Eyre, D.and Miller, H.G., *Sturmian projection and an L2 discretization of three-body continuum effects*, Phys Rev C: Nucl Phys, Vol 32 (3), p 727-37, 1985.

115. Eyre, D. and Miller, H.G., *Sturmian approximation of three-body continuum effects*, Phys Lett B, Vol 153B (1-2), p 5-7, 1985.

116. Eyre, D. and Miller, H.G., *Sturmian expansion approximation to three-body scattering*, Phys Lett B, Vol 129B (1-2), p 15-17, 1983.

117. Fano, Ugo, *Wave propagation and diffraction on a potential ridge*, Phys Rev Vol A 22, p 2660, 1980.

118. Fano, Ugo, *Unified treatment of collisions*, Phys Rev Vol A 24, p 2402, 1981.

119. Fano, Ugo, *Correlations of two excited electrons*, Rep Prog Phys Vol 46, p 97, 1983.

120. Fano, Ugo and Rao, A.R.P., **Atomic Collisions and Spectra**, Academic Press, Orlando, Florida, 1986.

121. Fernández Rico, J., Ramírez, G., López, R., and Fernández Alonso, J.I., Collect. Czech. Chem. Comm., Vol 53, p 2250, 1987.

122. Fernández Rico, J., López, R., Ema, I., and Ramírez, G., preprints, 1997.

123. Flores, J.C., *Kicked quantum rotator with dynamic disorder: a diffusive behavior in momentum space*, Phys Rev A, Vol 44 (6), p 3492-5, 1991.

124. Fock, V.A., Z. Phys., Vol 98, p 145, 1935.

125. Fock, V.A., *Hydrogen atoms and non-Euclidian geometry*, Kgl Norske Videnskab Forh, Vol 31, p 138, 1958.

126. Fonseca, A.C. and Pena, M.T., *Rotational-invariant Sturmian-Faddeev ansatz for the solution of hydrogen molecular ion (H2+): a general approach to molecular three- body problems*, Phys Rev A: Gen Phys, Vol 38 (10), p 4967-84, 1988.

127. Fonseca, A.C., *Four-body equations in momentum space*, Lect Notes Phys, Vol 273 (Models Methods Few-Body Phys.), p 161-200, 1987.

128. Fripiat, J.G.; Delhalle, J. and Defranceschi, M., *A momentum space approach to improve ab initio Hartree-Fock results based on the LCAO-GTF approximation*, NATO ASI Ser, Ser C, Vol 271 (Numer. Determ. Electron. Struct. At., Diat. Polyat. Mol.), p 263-8, 1989.

129. Gadre, Shridhar R. and Bendale, Rajeev D., *Maximization of atomic information-entropy sum in configuration and momentum spaces*, Int J Quantum Chem, Vol 28 (2), p 311-14, 1985.

130. Gadre, Shridhar R. and Chakravorty, Subhas, *The self-interaction correction to the local spin density model: effect on atomic momentum space properties*, Chem Phys Lett, Vol 120 (1), p 101-5, 1985.

131. Gallaher, D.F. and Wilets, L., *Coupled-state calculations of proton-hydrogen scattering in the Sturmian representation*, Phys Rev, Vol 169 (1), p 139-49, 1968.

132. Gazeau, J.P. and Maquet, A., *A new approach to the two-particle Schrödinger bound state problem*, J Chem Phys, Vol 73 (10), p 5147-54, 1980.

133. Gazeau, J.P. and Maquet, A., *Bound states in a Yukawa potential: a Sturmian group theoretical approach*, Phys Rev A, Vol 20 (3), p 727-39, 1979.

134. Geller, M., *Two-center Coulomb integrals*, J Chem Phys, Vol 41, p 4006, 1964.

135. Gerry, Christopher C., *Inner-shell bound-bound transitions from variationally scaled Sturmian functions*, Phys Rev A: Gen Phys, Vol 38 (7), p 3764-5, 1988.

136. Ghosh, Swapan K., *Quantum chemistry in phase space: some current trends*, Proc - Indian Acad Sci, Chem Sci, Vol 99 (1-2), p 21-8, 1987.

137. Gloeckle W., *Few-body equations and their solutions in momentum space*, Lect Notes Phys, Vol 273 (Models Methods Few-Body Phys.), p 3-52, 1987.

138. Goscinski, O., *Preliminary Research Report No. 217*, Quantum Chemistry Group, Uppsala University, 1968.

139. Gradshteyn, I.S. and Ryshik, I.M., **Tables of Integrals, Series and Products**, Academic Press, New York, (1965).

140. Grant, I.P., in **Relativistic Effects in Atoms and Molecules**, Wilson, S., Ed., Plenum Press, 1988.

141. Grant, I.P., in **Atomic, Molecular and Optical Physics Handbook**, Drake, G.W.F. Ed., Chapt 22, p 287, AIP Press, Woodbury New York, 1996.

142. Gruzdev, P.F.; Soloveva, G.S. and Sherstyuk, A.I., *Calculation of neon and argon steady-state polarizabilities by the method of Hartree-Fock SCF Sturmian expansion*, Opt Spektrosk, Vol 63 (6), p 1394-7, 1987.

143. Haftel, M.I. and Mandelzweig, V.B., *A fast convergent hyperspherical expansion for the helium ground state*, Phys Letters, Vol A 120, p 232, 1987.

144. Han, C.S., *Electron-atom scattering in an intense radiation field*, Phys Rev A: At, Mol, Opt Phys, Vol 51 (6), p 4818-23, 1995.

145. Hansen, T.B., *The many-center one-electron problem in momentum space*, Thesis, Chemical Institute, University of Copenhagen, 1998.

146. Harris, F.E. and Michels, H.H., Adv. Chem. Phys. **13**, 205, 1967.

147. Hartt, K. and Yidana, P.V.A., *Analytic Sturmian functions and convergence of separable expansions*, Phys Rev C: Nucl Phys, Vol 36 (2), p 475-84, 1987.

148. Heddle, David P; Kwon, Yong Rae and Tabakin, F., *Coulomb plus strong interaction bound states-momentum space numerical solutions*, Comput Phys Commun, Vol 38 (1), p 71-82, 1985.

149. Heller, E.J., *Wigner phase space method: Analysis for semiclassical applications*, J Chem Phys, Vol 65, 1289-1298, 1976.

150. Henderson, George A., *Variational theorems for the single-particle probability density and density matrix in momentum space*, Phys Rev A, Vol 23 (1), p 19-20, 1981.

151. Henriksen, N.E., Billing, G.D. and Hansen, F.Y., *Phase-space representation of quantum mechanics: Dynamics of the Morse oscillator*, Chem Phys Letters, Vol 148, 397-403.

152. Herrick, D.R., *Variable dimensionality in the group-theoretic prediction of configuration mixings for doubly-excited helium*, J Math Phys, Vol 16, p 1046, 1975.

153. Herrick, D.R., *New symmetry properties of atoms and molecules*, Adv Chem Phys, Vol 52, p 1, 1983.

154. Herschbach, Dudley R., *Dimensional interpolation for two-electron atoms*, J Chem Phys, Vol 84, p 838, 1986.

155. Herschbach, Dudley R., Avery, John and Goscinski, Osvaldo, Eds., **Dimensional Scaling in Chemical Physics**, Kluwer, Dordrecht, 1993.

156. Hietschold, M.; Wonn, H. and Renz, G., *Hartree-Fock-Slater exchange for anisotropic occupation in momentum space*, Czech J Phys, Vol B 35 (2), p 168-75, 1985.

157. Hillery, M., O'Connell, R.F., Scully, M.O., and Wigner, E.P., *Distribution functions in physics: Fundementals*, Physics Reports, Vol 106, 121-167, 1984.

158. Holoeien, E. and Midtdal, J., *Variational nonrelativistic calculations for the (2pnp)1,3Pe states of two-electron atomic systems*, J Phys B, Vol 4 (10), p 1243-9, 1971.

159. Holz, J., *Self-energy of electrons in a Coulomb field: momentum-space method*, Z Phys D: At, Mol Clusters, Vol 4 (3), p 211-25, 1987.

160. Horacek, Jiri and Zejda, Ladislav, *Sturmian functions for nonlocal interactions*, Czech J Phys, Vol 43 (12), p 1191-201, 1993.

161. Hughs, J.W.B., Proc Phys Soc, Vol 91, p 810, 1967.

162. Hua, L.K., **Harmonic Analysis of Functions of Several Complex Variables in the Classical Domains**, American Mathematical Society, Providence, R.I., 1963.

163. Ihm, J.; Zunger, Alex and Cohen, Marvin L., *Momentum-space formalism for the total energy of solids*, J Phys C, Vol 12 (21), p 4409-22, 1979.

164. Ishikawa, Yasuyuki; Rodriguez, Wilfredo and Alexander, S.A., *Solution of the integral Dirac equation in momentum space*, Int J Quantum Chem, Quantum Chem Symp, Vol 21, p 417-23, 1987.

165. Ishikawa, Yasuyuki; Rodriguez, Wilfredo; Torres, Samuel and Alexander S.A., *Solving the Dirac equation in momentum space: a numerical study of hydrogen diatomic monopositive ion*, Chem Phys Lett, Vol 143 (3), p 289-92, 1988.

166. Jain, Ashok and Winter, Thomas G., *Electron transfer, target excitation, and ionization in H+ + Na(3s) and H+ + Na(3p) collisions in the coupled-Sturmian-pseudostate approach*, Phys Rev A: At, Mol, Opt Phys, Vol 51 (4), p 2963-73, 1995.

167. Jain, Babu L., *A numerical study on the choice of basis sets used for translating ETOs in multi-center LCAO calculations*, ETO Multicent Mol Integr, Proc Int Conf, 1st, Reidel, Dordrecht, Neth, p 129-33, 81, Ed. Weatherford, Charles A. ; Jones, Herbert W., 1982.

168. Jasperse, J.R., *Method for one particle bound to two identical fixed centers: application to H_2^+*, Phys Rev A, Vol (3)2 (6), p 2232-44, 1970.

169. Jolicard, Georges and Billing, Gert Due, *Energy dependence study of vibrational inelastic collisions using the wave operator theory and an analysis of quantum flows in momentum space*, Chem Phys, Vol 149 (3), p 261-73, 1991.

170. Judd, B.R., **Angular Momentum Theory for Diatomic Molecules**, Academic Press, New York, 1975.

171. Kaijser, Per and Lindner, Peter, *Momentum distribution of diatomic molecules*, Philos Mag, Vol 31 (4), p 871-82, 1975.

172. Kaijser, Per and Sabin, John R., *A comparison between the LCAO-X.alpha. and Hartree-Fock wave functions for momentum space properties of ammonia*, J Chem Phys, Vol 74 (1), p 559-63, 1981.

173. Karule, E. and Pratt, R.H., *Transformed Coulomb Green function Sturmian expansion*, J Phys B: At, Mol Opt Phys, Vol 24 (7), p 1585-91, 1991.

174. Katyurin, S.V. and Glinkin, O.G., *Variation-iteration method for one-dimensional two-electron systems*, Int J Quantum Chem, Vol 43 (2), p 251-8, 1992.

175. Kellman, M.E. and Herrick, D.R., *Ro-vibrational collective interpretation of supermultiplet classifications of intrashell levels of two-electron atoms*, Phys Rev A, Vol 22, p1536, 1980.

176. Kil'dyushov, M.S., Sov J Nucl Phys, Vol 15, p 113, 1972.

177. Kil'dyushov, M.S., and Kuznetsov, G.I., Sov. J. Nucl. Phys., Vol 17, p 1330, 1973.

178. King, H.F., Stanton, R.E., Kim, H., Wyatt, R.E., and Parr, R.G., J. Chem. Phys., Vol 47, p 1936, 1967.

179. Klar, H., J Phys B, Vol 7, L436, 1974.

180. Klar, H. and Klar, M., *An accurate treatment of two-elecctron systems*, J Phys B, Vol 13, p 1057, 1980.

181. Klar, H., *Exact atomic wave functions - a generalized power-series expansion using hyperspherical coordinates*, J Phys A, Vol 18, p 1561, 1985.

182. Klarsfeld, S. and Maquet, A., *Analytic continuation of sturmian expansions for two-photon ionization*, Phys Lett A, Vol 73A (2), p 100-2, 1979.

183. Klarsfeld, S. and Maquet, A., *Pade-Sturmian approach to multiphoton ionization in hydrogenlike atoms*, Phys Lett A, Vol 78A (1), p 40-2, 1980.

184. Klepikov, N.P., Sov. J. Nucl. Phys. Vol 19, p 462, 1974.

185. Knirk, D.L., *Approach to the description of atoms using hyperspherical coordinates*, J Chem Phys, Vol 60, p 1, 1974.

186. Koga, Toshikatsu and Murai, Takeshi, *Energy-density relations in momentum space. III. Variational aspect*, Theor Chim Acta, Vol 65 (4), p 311-16, 1984.

187. Koga, Toshikatsu, *Direct solution of the $H(1s) - H^+$ long-range interactionproblem in momentum space*, J Chem Phys, Vol 82, p 2022, 1985.

188. Koga, Toshikatsu and Matsumoto, S., *An exact solution of the interaction problem between two ground-state hydrogen atoms*, J Chem Phys, Vol 82, p 5127, 1985.

189. Koga, Toshikatsu and Kawaai, Ryousei, *One-electron diatomics in momentum space. II. Second and third iterated LCAO solutions* J Chem Phys, Vol 84 (10), p 5651-4, 1986.

190. Koga, Toshikatsu and Matsuhashi, Toshiyuki, *One-electron diatomics in momentum space. III. Nonvariational method for single-center expansion*, J Chem Phys, Vol 87 (3), p 1677-80, 1987.

191. Koga, Toshikatsu and Matsuhashi, Toshiyuki, *Sum rules for nuclear attraction integrals over hydrogenic orbitals*, J Chem Phys, Vol 87 (8), p 4696-9, 1987.

192. Koga, Toshikatsu and Matsuhashi, Toshiyuki, *One-electron diatomics in momentum space. V. Nonvariational LCAO approaaach*, J Chem Phys, Vol 89, p 983, 1988.

193. Koga, Toshikatsu; Yamamoto, Yoshiaki and Matsuhashi, Toshiyuki, *One-electron diatomics in momentum space. IV. Floating single-center expansion*, J Chem Phys, Vol 88 (10), p 6675-6, 1988.

194. Koga, Toshikatsu and Ougihara, Tsutomu, *One-electron diatomics in momentum space. VI. Nonvariational approach to excited states*, J Chem Phys, Vol 91 (2), p 1092-5, 1989.

195. Koga, Toshikatsu; Horiguchi, Takehide and Ishikawa, Yasuyuki, *One-electron diatomics in momentum space. VII. Nonvariational approach to ground and excited states of heteronuclear systems*, J Chem Phys, Vol 95 (2), p 1086-9, 1991.

196. Kolos, W. and Wolniewicz, L., J. Chem. Phys., Vol 41, p 3663, 1964.

197. Kolos, W. and Wolniewicz, L., J. Chem. Phys., Vol 49, p 404, 1968.

198. Kramer, Paul J. and Chen, Joseph C.Y., *Faddeev equations for atomic problems. IV. Convergence of the separable-expansion method for low-energy positron-hydrogen problems*, Phys Rev A, Vol (3)3 (2), p 568-73, 1971.

199. Krause, Jeffrey L. and Berry, R. Stephen, *Electron correlation in alkaline earth atoms*, Phys Rev A, Vol 31 (5), p 3502-4, 1985.

200. Kristoffel, Nikolai, *Statistics with arbitrary maximal allowed number of particles in the cell of the momentum space (methodical note)*, Eesti Tead Akad Toim, Fuus, Mat, Vol 41 (3), p 207-10, 1992.

201. Kupperman, A. and Hypes, P.G., *3-dimensional quantum mechanical reactive scattering using symmetrized hyperspherical coordinates*, J Chem Phys, Vol 84, 5962, 1986.

202. Kuznetsov, G.I. and Smorodinskii, Ya., Sov. J. Nucl. Phys., Vol 25, p 447, 1976.

203. Lakshmanan, M. and Hasegawa, H., *On the canonical equivalence of the Kepler problem in coordinate and momentum spaces*, J Phys A: Math Gen, Vol 17 (16), 1984.

204. Landau, L.D., and Lifshitz, E.M., **Quantum Mechanics; Non-Relativistic Theory**, Pergamon Press, London, 1959.

205. Lassettre, Edwin N., *Momentum eigenfunctions in the complex momentum plane. V. Analytic behavior of the Schrödinger equation in the complex momentum plane. The Yukawa potential*, J Chem Phys, Vol 82 (2), p 827-40, 1985.

206. Lassettre, Edwin N., *Momentum eigenfunctions in the complex momentum plane. VI. A local potential function*, J Chem Phys, Vol 83 (4), p 1709-21, 1985.

207. Lin, C.D., *Analytical channel functions for 2-electron atoms in hyperspherical coordinates*, Phys. Rev. A, Vol 23, p 1585, 1981.

208. Linderberg, J. and Öhrn, Y., *Kinetic energy functional in hyperspherical coordinates*, Int J Quant Chem, Vol 27, p 273, 1985.

209. Liu, F.Q.; Hou, X.J. and Lim, T.K., *Faddeev-Yakubovsky theory for four-body systems with three-body forces and its one-dimensional integral equations from the hyperspherical-harmonics expansion in momentum space*, Few-Body Syst, Vol 4 (2), p 89-101, 1988.

210. Liu, F.Q. and Lim, T.K., *The hyperspherical-harmonics expansion method and the integral-equation approach to solving the few-body problem in momentum space*, Few-Body Syst, Vol 5 (1), p 31-43, 1988.

211. Lizengevich, A.I., *Momentum correlations in a system of interacting particles*, Ukr Fiz Zh (Russ Ed), Vol 33 (10), p 1588-91, 1988.

212. López, R., Ramírez, G., García de la Vega, J.M., and Fernández Rico, J., J Chim Phys, Vol 84, p 695, 1987.

213. Louck, J.D., *Generalized orbital angular momentum and the n-fold degenerate quantum mechanical oscillator*, J Mol Spectr, Vol 4, p 298, 1960.

214. Louck, J.D. and Galbraith, H.W., Rev Mod Phys, Vol 44, p 540, 1972.

215. Löwdin, P.O., Phys Rev Vol 97, p 1474, 1955.

216. Löwdin, P.O., Appl Phys Suppl, Vol 33, p 251, 1962.

217. McWeeny, Roy and Coulson, Charles A., *The computation of wave functions in momentum space. I. The helium atom*, Proc Phys Soc (London) A, Vol 62, p 509, 1949.

218. McWeeny, Roy, *The computation of wave functions in momentum space. II. The hydrogen molecule ion*, Proc Phys Soc (London) A, Vol 62, p 509, 1949.

219. Manakov, N.L.; Rapoport, L.P. and Zapryagaev, S.A., *Sturmian expansions of the relativistic Coulomb Green function*, Phys Lett A, Vol 43 (2), p 139-40, 1973.

220. Maquet, Alfred; Martin, Philippe and Veniard, Valerie, *On the Coulomb Sturmian basis*, NATO ASI Ser, Ser C, Vol 271 (Numer. Determ. Electron. Struct. At., Diat. Polyat. Mol.), p 295-9, 1989.

221. Maruani, Jean, editor, **Molecules in Physics, Chemistry and Biology, Vol 3**, Kluwer Academic Publishers, Dordrecht, 1989.

222. McCarthy, I.E. and Rossi, A.M., *Momentum-space calculation of electron-molecule scattering*, Phys Rev A: At, Mol, Opt Phys, Vol 49 (6), p 4645-52, 1994.

223. McCarthy, I.E. and Stelbovics, A.T., *The momentum-space coupled-channels-optical method for electron-atom scattering*, Flinders Univ South Aust, Inst At Stud, (Tech Rep) FIAS-R, (FIAS- R-111,), p 51 pp., 1983.

224. Michels, M.A.J., Int. J. Quantum Chem., Vol 20, p 951, 1981.

225. Mizuno, J., *Use of the Sturmian function for the calculation of the third harmonic generation coefficient of the hydrogen atom*, J Phys B, Vol 5 (6), p 1149-54, 1972.

226. Monkhorst, Hendrik J. and Harris, Frank E., *Accurate calculation of Fourier transform of two-center Slater orbital products*, Int J Quantum Chem, Vol 6, p 601, 1972.

227. Monkhorst, Hendrik J. and Jeziorski, Bogumil, *No linear dependence or many-center integral problems in momentum space quantum chemistry*, J Chem Phys, Vol 71 (12), p 5268-9, 1979.

228. Moore, C.E., **Atomic Energy Levels; Circular of the National Bureau of Standards 467**, Superintendent of Documents, U.S. Government Printing Office, Washington 25 D.C., 1949.

229. Navasa, J. and Tsoucaris, G., *Molecular wave functions in momentum space*, Phys Rev A, Vol 24, p 683, 1981.

230. Nikiforov, A.F., Suslov, S.K., and Uvarov, V.B., **Classical Orthogonal Polynomials of a Discrete Variable**, Springer-Verlag, Berlin, 1991.

231. Norbury, John W.; Maung, Khin Maung and Kahana, David E., *Exact numerical solution of the spinless Salpeter equation for the Coulomb potential in momentum space*, Phys Rev A: At, Mol, Opt Phys, Vol 50 (5), p 3609-13, 1994.

232. Novosadov, B.K., Opt Spectrosc, Vol 41, p 490, 1976.

233. Novosadov, B.K., Int J Quantum Chem, Vol 24, p 1, 1983.

234. Ojha, P.C., *The Jacobi-matrix method in parabolic coordinates: expansion of Coulomb functions in parabolic Sturmians*, J Math Phys (N Y), Vol 28 (2), p 392-6, 1987.

235. Park, Il Hung; Kim, Hong Ju and Kang, Ju Sang, *Computer simulation of quantum mechanical scattering in coordinate and momentum space* Sae Mulli, Vol 26 (4), p 155-67, 1986.

236. Pathak, Rajeev K; Kulkarni, Sudhir A. and Gadre, Shridhar R., *Momentum space atomic first-order density matrixes and "exchange-only" correlation factors*, Phys Rev A, Vol 42 (5), p 2622-6, 1990.

237. Pathak, Rajeev K; Panat, Padmakar V. and Gadre, Shridhar R., *Local-density-functional model for atoms in momentum space*, Phys Rev A, Vol 26 (6), p 3073-7, 1982.

238. Pauling, L., and Wilson, E.B., **Introduction to Quantum Mechanics**, McGraw-Hill, 1935.

239. Pisani, L. and Clementi, E., in **Methods and Techniques in Computational Chemistry**, Clemennti, E., and Corongiu, G., Eds., STEF, Cagliari, 1995.

240. Plante, D.R., Johnson, W.R., and Sapirstein, J., Phys. Rev. Vol A49, p 3519, 1994.

241. Podolski, B., Proc. Natl. Acad. Sci. (USA), Vol 14, p 253, 1928.

242. Podolski, B. and Pauling, L., Phys Rev, Vol 34, p 109, 1929.

243. Potvliege, R.M. and Shakeshaft, Robin, *Determination of the scattering matrix by use of the Sturmian representation of the wave function: choice of basis wave number*, J Phys B: At, Mol Opt Phys, Vol 21 (21), p L645, 1988.

244. Potvliege, R.M. and Smith, Philip H.G., *Stabilization of excited states and harmonic generation: Recent theoretical results in the Sturmian-Floquet approach*, NATO ASI Ser, Ser B, Vol 316 (Super-Intense Laser-Atom Physics), p 173-84, 1993.

245. Pyykkö, P., *Relativistic Theory of Atoms and Molecules. A Bibliography, 1916-1985*, Lecture Notes in Chemistry, Vol 41, 1986.

246. Pyykkö, P., Chem. Rev., Vol. 88, p 563, 1988.

247. Rahman, N.K., *On the Sturmian representation of the Coulomb Green's function in perturbation calculation*, J Chem Phys, Vol 67 (4), p 1684-5, 1977.

248. Rawitscher, G.H. and Delic, G., *Sturmian representation of the optical model potential due to coupling to inelastic channels*, Phys Rev C, Vol 29 (4), p 1153-62, 1984.

249. Rawitscher, George H. and Delic, George, *Solution of the scattering T matrix equation in discrete complex momentum space*, Phys Rev C, Vol 29 (3), p 747-54, 1984.

250. Regier, Philip E.; Fisher, Jacob; Sharma, B.S. and Thakkar, Ajit J., *Gaussian vs. Slater representations of d orbitals: An information theoretic appraisal based on both position and momentum space properties*, Int J Quantum Chem, Vol 28 (4), p 429-49, 1985.

251. Ritchie, Burke, *Comment on "Electron molecule scattering in momentum space"*, J Chem Phys, Vol 72 (2), p 1420-1, 1980.

252. Ritchie, Burke, *Electron-molecule scattering in momentum space*, J Chem Phys, Vol 70 (6), p 2663-9, 1979.

253. Rodriguez, Wilfredo and Ishikawa, Yasuyuki, *Fully numerical solutions of the Hartree-Fock equation in momentum space: a numerical study of the helium atom and the hydrogen diatomic monopositive ion*, Int J Quantum Chem, Quantum Chem Symp, Vol 22, p 445-56, 1988.

254. Rodriguez, Wilfredo and Ishikawa, Yasuyuki, *Fully numerical solutions of the molecular Schrödinger equation in momentum space*, Chem Phys Lett, Vol 146 (6), p 515-17, 1988.

255. Rohwedder, Bernd and Englert, Berthold Georg, *Semiclassical quantization in momentum space*, Phys Rev A: At, Mol, Opt Phys, Vol 49 (4), p 2340-6, 1994.

256. Rotenberg, Manuel, Ann. Phys. (New York), Vol 19, p 262, 1962.

257. Rotenberg, Manuel, *Theory and application of Sturmian functions*, Adv. At. Mol. Phys., Vol 6, p 233-68, 1970.

258. Royer, A., *Wigner function as the expectation value of a parity operator*, Phys Rev A, Vol 15, p 449-450, 1977.

259. Rudin, W., **Fourier Analysis on Groups**, Interscience, New York, 1962.

260. Schmider, Hartmut; Smith, Vedene H. Jr. and Weyrich, Wolf, *On the inference of the one-particle density matrix from position and momentum-space form factors*, Z Naturforsch, A: Phys Sci, Vol 48 (1-2), p 211-20, 1993.

261. Schmider, Hartmut; Smith, Vedene H. Jr. and Weyrich, Wolf, *Reconstruction of the one-particle density matrix from expectation values in position and momentum space*, J Chem Phys, Vol 96 (12), p 8986-94, 1992.

262. Schuch, Dieter, *On a form of nonlinear dissipative wave mechanics valid in position- and momentum-space*, Int J Quantum Chem, Quantum Chem Symp, Vol 28 (Proceedings of the International Symposium on Atomic, Molecular, and Condensed Matter Theory and Computational Methods, 1994), p 251-9, 1994.

263. Shabaev, V.M., *Relativistic Coulomb Green function with regard to finite size of the nucleus*, Vestn Leningr Univ, Fiz, Khim (2), p 92-6, 1984.

264. Shakeshaft, Robin and Tang, X., *Determination of the scattering matrix by use of the Sturmian representation of the wave function*, Phys Rev A: Gen Phys, Vol 35 (9), p 3945-8, 1987.

265. Shakeshaft, Robin, *A note on the Sturmian expansion of the Coulomb Green's function*. J Phys B: At Mol Phys, Vol 18 (17), p L611-L615, 1985.

266. Shakeshaft, Robin, *Application of the Sturmian expansion to multiphoton absorption: hydrogen above the ionization threshold*, Phys Rev A: Gen Phys, Vol 34 (6), p 5119-22, 1986.

267. Shakeshaft, Robin, *Coupled-state calculations of proton-hydrogen-atom scattering with a Sturmian expansion*, Phys Rev A, Vol 14 (5), p 1626-33, 1976.

268. Shakeshaft, Robin, *Sturmian expansion of Green's function and its application to multiphoton ionization of hydrogen*, Phys Rev A: Gen Phys, Vol 34 (1), p 244-52, 1986.

269. Shakeshaft, Robin, *Sturmian basis functions in the coupled state impact parameter method for hydrogen(+) + atomic hydrogen scattering*, J Phys B, Vol 8 (7), p 1114-28, 1975.

270. Shelton, D.P., *Hyperpolarizability of the hydrogen atom*, Phys Rev A: Gen Phys, Vol 36 (7), p 3032-41, 1987.

271. Sherstyuk, A.I., *Sturmian expansions in the many-fermion problem*, Teor Mat Fiz, Vol 56 (2), p 272-87, 1983.

272. Shibuya, T. and Wulfman, C.E., *Molecular orbitals in momentum space*, Proc Roy Soc A, Vol 286, p 376, 1965.

273. Shull, H. and Löwdin, P.-O., *Superposition of configurations and natural spin-orbitals. Applications to the He problem*, J Chem Phys, Vol 30, p 617, 1959

274. Simas, Alfredo M.; Thakkar, Ajit J. and Smith, Vedene H. Jr., *Momentum space properties of various orbital basis sets used in quantum chemical calculations*, Int J Quantum Chem, Vol 21 (2), p 419-29, 1982.

275. Sloan, I.H. and Gray, J.D., *Separable expansions of the t-matrix*, Phys Lett B, Vol 44 (4), p 354-6, 1973.

276. Sloan, Ian H., *Sturmian expansion of the Coulomb t matrix*, Phys Rev A, Vol 7 (3), p 1016-23, 1973.

277. Smirnov, Yu. F. and Shitikova, K.V., Sov J Part Nucl, Vol 8, p344, 1976.

278. Smith, F.T., *Generalized angular momentum in many-body collisions*, Phys Rev, Vol 120, p 1058, 1960.

279. Smith, F.T., *A symmetric representation for three-body problems. I. Motion in a plane*, J Math Phys, Vol 3, p 735, 1962.

280. Smith, F.T., *Participation of vibration in exchange reactions*, J Chem Phys, Vol 31, p 1352-1359, 1959.

281. Smith, Vedene H. Jr., *Density functional theory and local potential approximations from momentum space considerations*, Local Density Approximations Quantum Chem Solid State Phys, (Proc Symp), Plenum, New York, N. Y, p 1-19, 82, Eds. Dahl, Jens Peder; Avery, John, 1984.

282. Smorodinskii, Ya., and Efros, V.D., Sov. J. Nucl. Phys. Vol 17, p 210, 1973.

283. Springborg, M. and Dahl, J.P., *Wigner's phase-space function and atomic structure*, Phys Rev A, Vol 36, p 1050-1062, 1987.

284. Szmytkowski, R., The Dirac-Coulomb Sturmians and the Series Expansion of the Dirac-Coulomb Green Function; Application to the Relativistic Polarizability of the Hydrogenlike Atom, J. Phys. A, Vol 31, p 4963, 1998.

285. Szmytkowski, R., The Continuum Schrödinger-Coulomb and Dirac-Coulomb Sturmian Functions, J. Phys. A, Vol 31, p 4963, 1998.

286. Szmytkowski, R., The Continuum Schrödinger-Coulomb and Dirac-Coulomb Sturmian Functions, J. Phys. A, Vol 31, p 4963, 1998.

287. Taieb, Richard; Veniard, Valerie; Maquet, Alfred; Vucic S. and Potvliege R.M., *Light polarization effects in laser-assisted electron-impact-ionization ((e,2e)) collisions: a Sturmian approach*, J Phys B: At, Mol Opt Phys, Vol 24 (14), p 3229-40, 1991.

288. Tang, X. and Shakeshaft, R., *A note on the solution of the Schrödinger equation in momentum space*, Z Phys D: At, Mol Clusters, Vol 6 (2), p 113-17, 1987.

289. Tarter, C.B., J Math Phys, Vol 11, p 3192, 1970.

290. Tel-nov, D.A., *The d.c. Stark effect in a hydrogen atom via Sturmian expansions*, J Phys B: At, Mol Opt Phys, Vol 22 (14), p L399-L404, 1989.

291. Thakkar, Ajit J. and Koga, Toshikatsu, *Analytic approximations to the momentum moments of neutral atoms*, Int J Quantum Chem, Quantum Chem Symp, Vol 26 (Proc. Int. Symp. At., Mol., Condens. Matter Theory Comput. Methods, 1992), p 291-8, 1992.

292. Thakkar, Ajit J. and Tatewaki, Hiroshi, *Momentum-space properties of nitrogen: improved configuration-interaction calculations*, Phys Rev A, Vol 42 (3), p 1336-45, 1990.

293. Tzara, C., *A study of the relativistic Coulomb problem in momentum space*, Phys Lett A, Vol 111A (7), p 343-8, 1985.

294. Ugalde, Jesus M., *Exchange-correlation effects in momentum space for atoms: an analysis of the isoelectronic series of lithium 2S and beryllium 1S*, J Phys B: At Mol Phys, Vol 20 (10), p 2153-63, 1987.

295. Van Haeringen, H. and Kok, L.P., *Inequalities for and zeros of the Coulomb T matrix in momentum space*, Few Body Probl Phys,

Proc Int IUPAP Conf, 10th, North-Holland, Amsterdam, Neth, p 361-2, 83, Ed. Zeitnitz, Bernhard, 1984.

296. Vilenkin, N.K., **Special Functions and the Theory of Group Representations**, American Mathematical Society, Proovidence, R.I., 1968.

297. Vilenkin, N. Ya.; Kuznetsov, G.I., and Smorodinskii, Ya.A., Sov J Nucl Phys, Vol 2, p 645, 1966.

298. Vladimirov, Yu.S. and Kislov, V.V., *Charge of the nucleus of a hydrogen-like atom as an eigenvalue of a 6-dimensional wave equation in momentum space*, Izv Vyssh Uchebn Zaved, Fiz, Vol 28 (4), p 66-9, 1985.

299. Weatherford, Charles A., *Scaled hydrogenic Sturmians as ETOs*, ETO Multicent Mol Integr, Proc Int Conf, 1st, Reidel, Dordrecht, Neth, p 29-34, 81, Ed. Weatherford, Charles A. ; Jones, Herbert W., 1982.

300. Wen, Zhen-Yi and Avery, John, *Some properties of hyperspherical harmonics*, J Math Phys, Vol 26, 396, 1985.

301. Weniger, E.J., *Weakly convergent expansions of a plane wave and their use in Fourier integrals*, J Math Phys, Vol 26, p 276, 1985.

302. Weniger, E.J. and Steinborn, E.O., *The Fourier transforms of some exponential-type basis functions and their relevance for multicenter problems*, J. Chem Phys, Vol 78, p 6121, 1983.

303. Weniger, E.J.; Grotendorst, J., and Steinborn, E.O., *Unified analytical treatment of overlap, two-center nuclear attraction, and Coulomb integrals of B functions via the Fourier transform method*, Phys Rev A, Vol 33, p 3688, 1986.

304. Whitten, R.C. and Sims, J.S., Phys Rev A, Vol 9, p 1586, 1974.

305. Wigner, E., Phys Rev, Vol 40, p 749, 1932.

306. Windt, Laurent de; Fischer, Patrick; Defranceschi, Mireille; Delhalle, Joseph and Fripiat, Joseph G., *A combined analytical and numerical strategy to solve the atomic Hartree-Fock equations in momentum space*, J Comput Phys, Vol 111 (2), p 266-74, 1994.

307. Winter, Thomas G. and Alston, Steven G., *Coupled-Sturmian and perturbative treatments of electron transfer and ionization in high-energy helium p-He+ collisions*, Phys Rev A, Vol 45 (3), p 1562-8, 1992.

308. Winter, Thomas G., *Electron transfer and ionization in collisions between protons and the ions lithium(2+) and helium(1+) studied with the use of a Sturmian basis*, Phys Rev A: Gen Phys, Vol 33 (6), p 3842-52, 1986.

309. Winter, Thomas G., *Coupled-Sturmian treatment of electron transfer and ionization in proton-neon collisions*, Phys Rev A, Vol 48 (5), p 3706-13, 1993.

310. Winter, Thomas G., *Electron transfer and ionization in collisions between protons and the ions helium(1+), lithium(2+), beryllium(3+), boron(4+), and carbon(5+) studied with the use of a Sturmian basis*, Phys Rev A: Gen Phys, Vol 35 (9), p 3799-809, 1987.

311. Winter, Thomas G., *Electron transfer in p-helium(1+) ion and helium(2+) ion-atomic helium collisions using a Sturmian basis*, Phys Rev A, Vol 25 (2), p 697-712, 1982.

312. Winter, Thomas G., *Sturmian treatment of excitation and ionization in high-energy proton-helium collisions*, Phys Rev A, Vol 43 (9), p 4727-35, 1991.

313. Winter, Thomas G., *Coupled-Sturmian treatment of electron transfer and ionization in proton-carbon collisions*, Phys Rev A, Vol 47 (1), p 264-72, 1993.

314. Winter, Thomas G., *Electron transfer and ionization in proton-helium collisions studied using a Sturmian basis*, Phys Rev A, Vol 44 (7), p 4353-67, 1991.

315. Wulfman, Carl E., *Semiquantitative united-atom treatment and the shape of triatomic molecules*, J Chem Phys Vol 31, p 381, 1959.

316. Wulfman, Carl E., *Dynamical groups in atomic and molecular physics*, in **Group Theory and its Applications**, Loebel, E.M. Ed., Academic Press, 1971.

317. Wulfman, Carl E., *Approximate dynamical symmetry of two-electron atoms*, Chem Phys Letters Vol 23 (3), 1973.

318. Wulfman, Carl E., *On the space of eigenvectors in molecular quantum mechanics*, Int J Quant Chem Vol 49, p 185, 1994.

319. Yurtsever, Ersin; Yilmaz, Osman and Shillady, D.D., *Sturmian basis matrix solution of vibrational potentials.* Chem Phys Lett, Vol 85 (1), p 111-16, 1982.

320. Yurtsever, Ersin, *Franck-Condon integrals over a sturmian basis. An application to photoelectron spectra of molecular hydrogen and molecular nitrogen*, Chem Phys Lett, Vol 91 (1), p 21-6, 1982.

Index